The
Science Times
Book of
MAMMALS

Other titles in the series

The
Science Times
Book of
MAMMALS

EDITED BY
NICHOLAS WADE

THE LYONS PRESS

Printed in the United States of America

Designed by Joel Friedlander Publishing Services, San Rafael, CA

FIRST EDITION

10 9 8 7 6 5 4 3 2 1

Library of Congress Cataloging-in-Publication Data

The Science times book of mammals / edited by Nicholas Wade.
    p. cm.
  ISBN 1-55821-892-0
  1. Mammals. I. Wade, Nicholas. II. Science times.
  QL703.S415 1999
  599—dc21   98-49798
                                                               CIP

Contents

Introduction

It's easy to divide the living world into us and them, humans and the rest. But if you walk around a zoo, another division quickly comes to mind. In some enclosures there are warm, soft, furry creatures with what seem to be bright expressions and intelligent eyes. And then there are other animals, cold and scaly, their faces set in masks of implacable indifference or hostility.

The animals we identify with are, of course, the mammals, the alien ones the reptiles. Mammals are creatures of the breast and of motherhood. Reptilian is the customary epithet for tempters of Eve and extraterrestrial invaders. Mammals are members of the household, whether as adoptees, like cats and dogs, or as uninvited guests, like rats, mice and raccoons.

Mammals seem so familiar because they are, in fact, our cousins, part of the extended family of animals that includes humans. However much we may differ from other mammals in our intelligence and culture, our body plan and basic behavioral patterns are very much like theirs.

Other mammals, even those with a fraction of our intellect, are so like us in some ways that we can read their thoughts, or think we can. When a cat yawns and stretches, or a dog cocks its head to one side, looking alternately at its owner and the door, their state of mind seems as clear as if they had announced it out loud.

The first mammals split off from the reptile line, from creatures known as therapsids, in the late Triassic some 230 to 208 million years ago. They seem to have been small and no doubt fairly helpless creatures, mere bystanders during the long dominance of the dinosaurs, a rival scion of the reptilian heritage.

During their understudy period, the mammals gradually acquired a suite of abilities that brought great advantages once the dinosaurs' day was done. The prototype mammals developed hair for insulation, to keep their bodies at the same temperature all the time. Hair or fur is still the hallmark of the mammalian body plan. (Whales have hair in the fetal stage, but quickly shed it.)

The inner ear is another distinguishing feature of mammals. By the process of natural selection, one of the bones of the reptilian jaw was reworked into the three little bones that transmit sound to the cochlea, endowing mammals with a new and sensitive hearing mechanism.

Keener hearing and vision were important in themselves, but brought about an even more significant change in body plan—a much larger and more complicated brain to handle the greater sensory input from ears and eyes. Mammals also developed the feature from which they take their name, the mammary gland, or breast, a specialized structure for feeding the young.

Equipped with intelligence and adaptability, mammals were better able than their reptilian forebears to take advantage of changing conditions. After their long nascence as small scurrying animals they branched out into a rich spectrum of creatures, adapted to all kinds of niches in the terrestrial world. No less than four orders of mammals, as a major grouping of species is termed, returned to the sea—the two kinds of whale, the dugongs and manatees, and several members of the order Carnivora, such as seals, sea lions and walruses. Another order, the bats, took to the air and attained all the dexterity of birds.

Some orders of mammal have already fallen extinct. Experts differ as to the number of existing orders, since there are different ways of grouping the various mammals. The following species exemplify the wide range of living mammalian families: the duck-billed platypus (monotremes); hedgehogs and moles (insectivores); elephant shrews (Macroscelidea); flying lemurs (Dermoptera); bats (Chiroptera); lemurs, monkeys, apes and humans (primates); sloths and armadillos (edentates); pangolins (Pholidota); aardvarks and anteaters (Tubulidentata); rabbits and hares (lagomorphs); rats, squirrels, beavers and porcupines (rodents); cats, dogs, bears, seals and walruses (Carnivora), dolphins and sperm whales (toothed whales); blue whales and right whales (baleen whales); elephants (Proboscidea); dugongs and manatees (Sirenia); hyraxes (hyracoids); rhinoceroses and horses (perissodactyls); pigs, cattle, sheep, antelopes and giraffes (artiodactyls).

Mammals are fascinating both in their own right and for the light they shed on human evolution. The following articles about mammals were written for the "Science Times" section of *The New York Times*. Brought together in book form, the articles provide an overview of mammals that have brought themselves to biologists' current attention or made the headlines for one reason or another.

My colleagues and I in the science section are grateful to Lilly Golden of the Lyons Press for the idea of this book, and for giving our ephemeral writings a second life.

1

FROM PLATYPUSES TO RHINOCEROSES

All mammals may be our close evolutionary cousins but, to talk within the family, some of us are really rather odd. The monotremes, now represented only by the platypus and the echidna (or spiny anteater), lay eggs, rather than give birth to live-born babies, and are armed with a venom-dispensing spur on their hind legs.

Sloths belong to a more evolved order of mammals but have some bizarre habits, such as hosting a large variety of moths and other species in their shaggy, uncombed hair.

Bats have more to be proud about, including the magnificent high-tech sonar with which they catch insect meals on the wing. But even the best weapons are eventually blunted by countermeasures. Some moths reflexively dive for the ground as soon they perceive the beep of the bat's sonar pulses.

Natural selection works within tight constraints: More of this means less of that. The high cost of crafting a special ability, like bat sonar, will be effective only if it gives the animal some edge in the battle for survival. So why is the pronghorn endowed with the ability to run like the wind when there is no predator that has a chance of catching it?

The articles that follow survey the wide range of shapes and styles in the mammalian fashion parade.

If a Platypus Had a Dream, What Would It Mean?

TO SLEEP, PERCHANCE to dream; but if so, what on Earth might a duck-billed platypus dream about?

Although scientists are not yet sure that the platypus does indeed dream, the animal spends a lot of time sleeping, and during much of that time it behaves as if it were dreaming. Tantalizing hints about its sleeping brain have emerged from a study conducted by American and Australian scientists.

The somnolent mind of the platypus is no joking matter. Investigation of platypus slumber may shed light on mammal evolution, the development of the brain, the function of sleep and the origin of dreams. It might even bear on the weighty question of whether or not dinosaurs had dreams.

The platypus, which hunts in rivers and lives in nearby burrows along Australia's east coast, is one of the strangest animals ever to evolve. Weighing about four pounds, it is a furry, warm-blooded creature with a rather low body temperature (89.6 degrees Fahrenheit) that lays eggs and nurses its hatchlings from slits in its abdominal wall. It has a rubbery beak, a flat, beaverlike tail and webbed feet, which, in the males of the species, are armed with a pair of poison-injecting spurs like the fangs of venomous snakes.

The platypus (formally called *Ornithorhynchus anatinus*) and two species of echidnas (or spiny anteaters) are the sole living species of the mammalian order Monotremata, known as the monotremes (meaning one-holers). Monotremes are widely regarded as the most primitive of all modern mammals, perhaps resembling some ancient creature that represented an evolutionary link between the reptiles and birds on one hand and the mammals on the other.

But much of what biologists surmised about monotremes over the years has turned out to be wrong. Until recently biologists assumed that monotremes, alone of all mammals, do not experience the type of sleep known as rapid eye movement, or REM, sleep. Now it begins to look as if the platypus, if not its evolutionary cousins, the echidnas, experiences at least some of the features of REM sleep after all.

REM sleep in human beings is associated with vivid dreaming, and its observable symptoms—eyes moving rapidly under closed lids, muscular twitching, rapid and irregular heartbeats and rapid, low-amplitude brain waves—are common to all mammals and birds, in varying degrees. The latest investigation of platypus REM sleep was conducted by Dr. Jerome M. Siegel, a professor of psychiatry at the University of California at Los Angeles, and his Australian colleagues at the University of Queensland in Brisbane.

Their work, partly supported by a grant from the National Institutes of Health and reported at a meeting of the Society for Neuroscience in New Orleans, showed that something goes on in the sleeping platypus brain that resembles REM sleep, although it is different in one important way.

"When you or I have an interlude of REM sleep," Dr. Siegel said in an interview, "an electroencephalographic trace of our brain waves will look as if we were awake. The brain stem, or pons, initiates the REM episode and the electrical patterns common to both the waking state and REM sleep are then taken up by the forebrain. That is where we detect the low-amplitude, fast-wave electrical emissions characteristic of REM sleep.

"But with the platypus, you don't see this electrical REM pattern in the forebrain, even when the animal is rapidly moving its closed eyes, twitching, and showing other signs of REM sleep," he said. "In the platypus we're seeing REM sleep originating in the brain stem but without the involvement of the forebrain."

Dr. Siegel sees an evolutionary clue in this.

Although all adult mammals produce REM-type electrical waves from their forebrains, infant mammals, including human babies, do not, he said. Although babies have lots of REM sleep, this state does not begin to show up in electrical signals in the forebrain until a baby has developed its sensory systems.

"I didn't expect to see this," Dr. Siegel said, "but it seems that the platypus's form of REM sleep, apparently involving only its brain stem and not its forebrain, is like a human baby's form of REM sleep."

Dr. Siegel said the human baby in its early development may mimic the primitive evolutionary state of its distant ancestors, in which the forebrain was not involved in REM sleep, a constraint that the modern platypus retains throughout its life. The theory that the evolutionary history of a species is replayed in the development of fetuses and babies is known as "ontogeny recapitulates phylogeny."

In fact, Dr. Siegel said he believed the REM brain activity that is confined to the pons evolved long before the forebrain got into the act—so long ago that it could easily have been experienced by the dinosaurs. "Our work indicates that dinosaurs may indeed have had REM sleep," he said. Whether or not this implies that they had dreams is an open (and probably unanswerable) question.

The embryo platypus itself goes through stages that may mirror its evolution; the fetus starts out with little teeth of the kind its reptile ancestors probably had, but as the fetus develops, the teeth disappear and are replaced by a large bill.

REM sleep is not necessarily an attribute of intelligence, even though this sleeping state seems to be related to the processing of information by the brain. Humans have only a moderate amount of REM sleep, Dr. Siegel said, and dolphins, generally considered to be intelligent mammals, have very little REM sleep. (In non-REM sleep, he said, a dolphin apparently shuts down only one hemisphere of its brain at a time, alternating from side to side. In that state the dolphin swims in a slow circle, and when the sleeping side of the brain switches, the animal reverses its circular course.)

Studying the platypus poses special problems, Dr. Siegel said. For one thing, handling a male platypus exposes a person to the animal's poison spur. The sting is not normally fatal to human beings, but it causes excruciating pain that can persist for months.

To observe platypuses in realistic conditions, scientists at the University of Queensland installed a special water tank for four of these semi-aquatic animals, and built a long burrow for them nearby. A video camera was installed in the burrow to observe eye movement, and sensors recording brain activity, heartbeat and other functions were attached to the animals. Radio transmitters relayed information from the sensors to receivers outside the burrow.

"Platypuses placed in zoos often die within a few months," Dr. Siegel said, "but all of ours flourished because we shielded them electrically."

The platypus is extremely sensitive to electric fields, he said, because it uses electroreceptors in its beak to seek prey. The platypus searches for worms and small shrimps with its eyes closed, using its beak as a scanner to track the faint electric signals emitted by the beating hearts of its prey.

But in a zoo or laboratory, the researchers discovered, platypus tanks are ordinarily exposed to large electric fields emitted by the filtration equipment used to clean the water. These electric fields apparently confuse and disorient the animals and disable their hunting mechanism.

The broad class of mammals consists of three orders, or branches: the monotremes; the marsupials, which nurture their young in pouches; and the placental mammals (including human beings), which give birth to off-spring immediately capable of living outside their mothers' bodies. Until recently, most biologists believed that the monotremes were the first mammals to evolve from reptiles, and that marsupials and placentals evolved as separate branches at some later time.

But new studies in Sweden, Germany and Australia of the mitochondrial DNA of these three mammalian lineages challenge this view by showing that monotremes were probably originally part of the marsupial lineage. (Mitochondrial DNA, genetic material passed from one generation to the next by mothers alone, changes with time. It is used by scientists as a kind of evolutionary clock.)

In a paper published in *The Proceedings of the National Academy of Sciences*, Dr. Axel Janke and his colleagues at the University of Lund, Sweden, reported their conclusions based on mammalian mitochondrial DNA. They found that the three mammalian lineages had a common ancestor before 130 million years ago, at which time the placentals split off. The mammals destined to evolve into monotremes and marsupials continued to share a common branch until they diverged from each other 115 million years ago, the Swedish scientists found.

Today there are about 4,000 placental species, 300 marsupial species and 3 monotreme species. (Two extinct monotreme species are also known from their fossils.)

REM sleep and dreaming appear to have evolved along with other physiological features.

Dreams were interpreted by the ancients as portents, by Sigmund Freud as cryptic representations of hidden thoughts and feelings, and by Dr. Francis Crick, codiscoverer of the structure of DNA, as a kind of eraser for wiping out excessive unneeded information from the brain.

Dr. Jonathan Winson, a specialist in dreaming and REM sleep at Rockefeller University, takes a different view. He believes that dreams reflect a fundamental aspect of mammalian memory processing—a need by the brain to mull over and organize what it has absorbed while awake.

In an article on the meaning of dreams in a special edition of *Scientific American,* Dr. Winson wrote that an echidna does not exhibit REM sleep—a fact that suggests REM sleep evolved as part of the sleep cycle when placental mammals branched off.

Dr. Winson said in an interview that Dr. Siegel's observations of a form of REM sleep in the platypus do not necessarily conflict with the view that echidnas are not known to experience REM sleep, even though they are all monotremes. "The two species may have split apart, creating a difference in their sleep behavior," he said.

A deeper question that Dr. Siegel and his Australian colleagues intend to pursue is whether or not reptiles experience REM sleep. The present consensus, he said, is that they do not, but this is uncertain.

"There have been reports of eye movement in sleeping chameleons and other reptiles," he said, "but the problem is, if you see behavior in a reptile that looks like REM sleep, it's hard to know whether the animal is really in REM sleep or is merely waking up." Moreover, he said, the brains of reptiles are organized so differently from those of mammals that it is difficult to identify correspondences in their neurological structures.

To better detect the REM state (if any) in reptiles and other animals, it will be necessary to probe specialized cells in the pons that are involved in initiating REM sleep, Dr. Siegel said.

"These cells contain norepinephrine, serotonin and other neurotransmitters known to have huge roles in mental health," he said. "Most of the antidepressant drugs in use these days are selective serotonin reuptake inhibitors, for example."

Such drugs, like Prozac, interfere with reabsorption of serotonin released by brain cells, thus raising levels of serotonin in the nerve synapses. Serotonin levels affect mood, sexual desire, digestion, depression and, apparently, REM sleep.

The platypus is an excellent subject for the investigation of sleep-related questions, partly because it spends so much of its time asleep, oblivious to the scientists observing it.

"We can bring a video camera up to within a few inches of its closed eyes in the artificial burrow without waking the animal," Dr. Siegel said. "The platypus has no natural enemies to speak of, and it is not an endangered species.

"Maybe that's why it sleeps so soundly and is such a good subject for us."

—MALCOLM W. BROWNE, December 1997

Legendary Giant Sloth Sought by Scientists in Amazon Rain Forest

AN AMERICAN BIOLOGIST and a team of scientists, technicians and Indian guides are preparing to penetrate the trackless rain forest of western Brazil in pursuit of a South American counterpart of the fabled Himalayan yeti.

The object of this quest, headed by Dr. David C. Oren, an American ornithologist employed by the Brazilian government, is an animal Dr. Oren

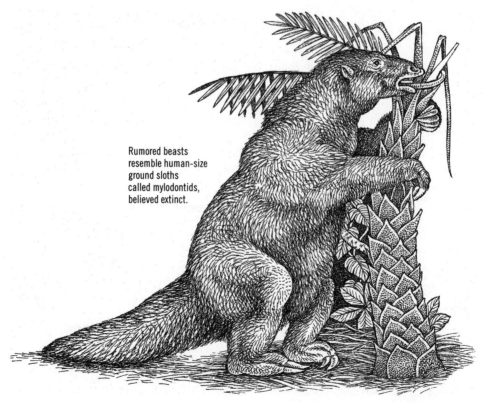

Rumored beasts resemble human-size ground sloths called mylodontids, believed extinct.

Patricia J. Wynne

believes to be a human-size ground sloth, belonging to a family thought by paleontologists to be long extinct.

Accounts by Indians of the Amazon region describe the elusive animal as terrifying and dangerous, physically powerful and equipped with some kind of chemical defense capable of paralyzing opponents. Dr. Oren, a staff scientist at the Goeldi Natural History Museum in Belém, Brazil, said by telephone that he had conducted more than 100 interviews in the last nine years with Indians and rubber tappers who told of having had contacts with the creature.

Dr. Oren acknowledges that he has had trouble persuading other scientists of the possibility that the creature is anything more than a local myth.

Among the American biologists he has sought to interest in the search is Dr. Malcolm C. McKenna, a paleontologist at the American Museum of Natural History in New York City. Dr. McKenna is dubious.

"While there is always a chance of discovering some previously unknown large animal somewhere in the world, just as it is possible to discover some new island in mid-ocean," Dr. McKenna said, "the likelihood is too small to draw me away from my work here. You can't just go chasing each rumor of a Sasquatch or yeti.

"On the other hand, discoveries of large new animals have sometimes surprised scientists: the vuquang ox, a large ungulate in postwar Vietnam; the okapi, discovered in Africa in 1900; a new peccary in the South American Chaco region believed to be extinct since the Pleistocene epoch. You never know."

But Dr. McKenna said it would take more than word-of-mouth accounts to convince scientists of the reality of a ground sloth living in the Amazon Basin. "I think scientists will insist on seeing at least a chunk of it," he said. "Even a photograph won't do."

Dr. Oren agrees that to convince skeptics he would have to bring the creature back, dead or alive.

"We'll be bringing tranquilizing dart guns," he said, "although it may be difficult to use them effectively." Most Indian accounts of the creature describe it as having an extremely tough skin that cannot be easily penetrated.

"That description tallies with fossil remains of a family of extinct ground sloths known as the mylodontids," he said. "These animals had dermal ossicles—bony armor plating embedded in the skin."

Reports have long circulated about a "mapinguari," a legendary and terrifying manlike creature of the vast Amazon rain forest, but when scientists took an interest at all, Dr. Oren said, they tended to guess that the animal, if it existed, might be some kind of primate.

"But when I began hearing accounts of a creature with shaggy red hair, backward-turned feet and a monkeylike face, I realized that witnesses might have encountered a ground sloth, closely related to extinct giant sloths known only from their fossils."

Three families of sloths are known, and only two genera, both of them tree dwellers, are known to have survived to the present day, Dr. Oren said. The common three-toed sloth is a lethargic creature that seems to live in slow motion, and is considered a family member of the megatheriids, known mostly from fossils.

The much rarer modern two-toed sloth, a surviving member of the megalonychid family, is less "slothful," Dr. Oren said, and can move swiftly and forcefully if threatened. The third family of sloths, the extinct, ground-dwelling mylodontids, grew to the size of large bears and were apparently very active.

The mapinguari, as Indians call the supposed forest creature, is apparently smaller than fossil members of the family, standing only about six feet tall when walking on its hind feet, but weighing some 500 pounds, and with jaws and feet powerful enough to rip palm trees apart. The creature is said to subsist largely on palm hearts and other vegetable delicacies of the rain forest.

The mapinguari is also described as having a thunderous voice that can sound quite human, and that has deceived human visitors to its habitat into thinking that another human was nearby.

Purported sightings of the creatures over a wide area generally agree on its appearance, Dr. Oren said, including descriptions of a ridge of manelike fur along the animal's neck and back.

"The Indians are very frightened of it, but some of them are anxious to capture one to prove to outsiders that it exists," Dr. Oren said. "Ten of these Indians will accompany our expedition."

Provided Dr. Oren and his group obtain necessary government permits in time, they expect to remain in the rain forest for about a month. To reach the interior of the state of Acre near the point where the frontiers of Brazil meet those of Peru and Bolivia—one of the regions where sightings have been reported—the group will travel to an Indian village, then upriver by boat, and then by foot, at least two days' march into the trackless forest. The creature is apparently never sighted along waterways where there is human habitation, but only in the depths of the forest.

A major motive for the search, Dr. Oren said, is to demonstrate that a large number of species in the Amazon Basin remain unknown to science but are nevertheless worthy of protection and conservation.

"My whole reason for being in the Amazon as a scientist," Dr. Oren said, "is to survey its enormous variety of species. If this ground sloth exists, it may be the largest land mammal in South America, and yet it is still unknown to science. If we find it, we will have proof that there are vast biological riches here that still await discovery."

—MALCOLM W. BROWNE, February 1994

Far from Fearsome, Bats Lose Ground to Ignorance and Greed

THE REPUTATION of bats leaves much to be desired. Creatures of caves and twilight, they evoke childhood fears of the dark and images of vampires. But real bats are not the aggressive, dirty or dangerous creatures of lore. They are, in fact, industrious and invaluable assets to people and the planet. And they are succumbing worldwide at a frightening rate to human ignorance, greed and destruction.

In misguided efforts to protect people, crops and livestock, bats are being poisoned, blasted with dynamite, asphyxiated by burning tires or entombed alive by the millions in caves and mines where they seek a day's or a winter's rest. Many others are felled by hunters or die when spelunkers disturb their dwellings.

"Bats are disappearing at a faster rate than any other group of vertebrates," said Dr. Merlin D. Tuttle, a real-life "batman" who has spearheaded international efforts to save the versatile and vital animals.

As scientists learn more about bats, they are discovering how vital they are to hundreds of environmentally and economically important plants and trees, including major features of the tropical rain forest, African savannah and American desert. And the demise of insect-eating bats has permitted a population explosion among insects that pester people and devour crops.

Bats, the only mammals that can fly, account for nearly a quarter of all known species of mammals. But 40 percent of bat species are listed as endangered or threatened. Several of the nearly 1,000 species have become extinct within the last two decades, some disappearing before scientists could determine their precise role in the chain of life.

Yet conservation organizations, conscious of the creature's charmless image, are loath to intervene.

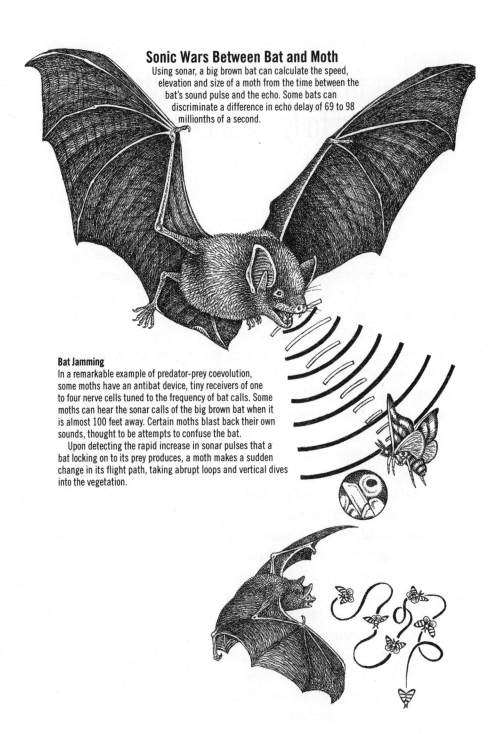

Sonic Wars Between Bat and Moth

Using sonar, a big brown bat can calculate the speed, elevation and size of a moth from the time between the bat's sound pulse and the echo. Some bats can discriminate a difference in echo delay of 69 to 98 millionths of a second.

Bat Jamming

In a remarkable example of predator-prey coevolution, some moths have an antibat device, tiny receivers of one to four nerve cells tuned to the frequency of bat calls. Some moths can hear the sonar calls of the big brown bat when it is almost 100 feet away. Certain moths blast back their own sounds, thought to be attempts to confuse the bat.

Upon detecting the rapid increase in sonar pulses that a bat locking on to its prey produces, a moth makes a sudden change in its flight path, taking abrupt loops and vertical dives into the vegetation.

Patricia J. Wynne

"When I first approached major conservation organizations 10 years ago, they wouldn't touch bats with a 10-foot pole," said the 50-year-old Dr. Tuttle, who since his high school days has been one of the furry animal's staunchest advocates. "They either didn't understand bats or viewed them as so unpopular as not to be worth the effort."

So Dr. Tuttle, then curator of mammals at the Milwaukee Public Museum, took bats literally and figuratively into his own hands. In 1982 he founded Bat Conservation International to sponsor research and educational programs about bats and their vital roles. Four years later, he left his museum job and moved the fledgling organization to Austin, Texas, a state that has the most bat species—32 of the nation's 42 kinds of bats—as well as the world's largest remaining bat colony, in Bracken Cave in central Texas, where 20 million Mexican free-tailed bats congregate.

Since then, Bat Conservation International has far surpassed its founder's dreams. Having arrived in Austin in 1986 with two employees and a few hundred members, it now boasts 17 full-time staff members and 12,500 members from 55 countries. Dr. Tuttle is the subject of a new biography for young people, *Batman: Exploring the World of Bats,* by Laurence Pringle (Macmillan), and his own book, *America's Neighborhood Bats*, is the University of Texas Press's bestselling volume.

A cartoon by W. Miller in *The New Yorker* is further testimony to Dr. Tuttle's mushrooming success. One of a group of bats hanging in a cave says to a roostmate: "We're starting to get media coverage. Pass it on."

As bats decline drastically throughout the world, scientists are scrambling to understand more about their unique biology, their critical ecological roles and their basic survival needs. At the 21st Annual North American Symposium on Bat Research in Austin, some 300 scientists gathered to exchange new findings and suggest ways of rescuing bats from their only serious natural enemy: people.

At one time, bats were vulnerable only during their nightly feeding frenzy and only to predators like owls with limited appetites. Now people attack them during the day in caves, trees and abandoned mines and under bridges and eaves. Bats are especially vulnerable to mass annihilation because most species congregate in huge groups when not eating.

Chased from most of their traditional roosts in caves, they moved into abandoned mines. Now thousands of mines are being sealed to prevent injury to curious people, and other mines are being reactivated or used for other purposes. In Wisconsin, for example, where 95 percent of the state's bats hibernate in old mines, Dr. Tuttle rescued hundreds of thousands of insect-eating bats by staying the hand of a landowner who was going to fumigate them so he could use the shaft to store cheese.

In tropical Central and South America, where the three species of blood-sipping vampire bats do an estimated $50 million worth of damage to livestock, ranchers who cannot tell one bat from another have destroyed millions of fruit-eating and insect-eating bats. Vampire bats, which roost in small groups, are far harder to find than the millions of insect eaters that gather in caves, where they can be smoked out by burning tires.

The critically important fruit-eating bats, however, have not fared well at human hands. In Southeast Asia, for example, giant fruit-eating bats known as flying foxes are routinely hunted for food and export, commanding as much as $20 apiece from wealthy islanders who consider them a gourmet treat. These huge bats, weighing as much as a pound and with wingspans of up to six feet, are easy targets for shotguns while they hang by the hundreds from treetops. Dr. Tuttle said that one of the Philippines' flying foxes, first discovered by scientists in the 1960s, was extinct by the 1980s.

Some 250 species of bats pollinate or disperse the seeds of hundreds of trees and shrubs, including those that bear economically important products like bananas, avocados, vanilla beans, dates, peaches, figs, cashews and agaves, the source of tequila. In a single night, one short-tailed fruit bat can disperse as many as 60,000 rain forest seeds.

A recent study by Bat Conservation International documented that more than 300 plant species in the Old World tropics alone rely on bats. These plants are the source of more than 450 commercial products, including medicines, timber, ornamentals, fiber and cordage, and dyes and tannins, as well as foods and drinks.

In Africa, fast-declining flying foxes are the only known seed dispersers for the iroko tree, the source of millions of dollars' worth of timber annually. The giant baobab tree, an ecologically critical feature of East African savannahs, depends on bats to pollinate its flowers, which open only at night. The fruit bats of West Africa disperse nearly all the seeds of "pioneer plants" that start forests growing again on cleared land.

To Donald W. Thomas, an ecologist at the University of Sherbrooke in Quebec, fruit bats are "the keystone species on many tropical islands," where they are the sole pollinators and seed dispersers of long-lived rain forest trees.

In studies in the Ivory Coast, Dr. Thomas showed that when the seeds of tropical fruits pass through the digestive tract of fruit bats, they nearly always germinate, but those seeds that simply fall to the ground from the trees mostly do not. Furthermore, he found by studying bat droppings that the animals dispersed most of the seeds far from the mother tree, which greatly enhanced a seedling's chances of taking hold.

Migrating nectar-eating bats from Mexico were recently shown to be a key to the survival of the giant cactus and agaves that are the distinguishing features of the Sonoran Desert. Dr. Theodore H. Fleming and colleagues from the University of Miami at Coral Gables found that lesser long-nosed bats, a federally listed endangered species, follow "nectar corridors"—a succession of blooming organ-pipe and saguaro cacti—as they head north into Arizona in spring and then follow a reverse succession of blooming agaves when they head south to Mexico in late fall.

With each foray into the nectar trough of a desert flower, the long-nosed bat emerges covered with pollen, which is deposited at the bat's next stop. Without these long-nosed bats, which are ideally constructed for their job as pollinators, the giant cactus and agave could die out.

Bats that eat insects find their night's meal by emitting high-frequency sounds and locating targets by the returning echoes. All six species of bats that live in and around New York City are insect-eating "microbats," some of which consume up to 3,000 insects a night. Although all can see, their sonar tells far more than eyes can about a potential meal on the wing in the dark.

The high-frequency sounds bats emit and the echoes they process give these voracious hunters detailed information about the movement, distance, speed, trajectory, size and shape of possible prey. Bat sonar can detect an object as thin as a human hair and only eight-hundredths of an inch long.

Dr. Nobuo Suga, a biologist at Washington University in St. Louis, studies electrical impulses generated by individual neurons in the auditory center of the pearl-size brain of the mustached bat. He has determined that by using pulses with different frequencies the animal derives very precise details about a potential meal. He also found that the bats use harmonics to avoid collisions with other bats while hunting in dense packs.

Dr. Tuttle noted that a half million pounds of insects typically succumb each night to large foraging colonies of insectivorous bats. The million or so Mexican free-tailed bats that roost under the Congress Avenue Bridge in downtown Austin eat 14 tons of insects a night from March to November, said Dr. Tuttle, who convinced the city not to destroy them but instead to turn the bat colony's nightly emergence into a tourist attraction.

Dr. James H. Fullard, a zoologist at the University of Toronto, noted that before bats evolved more than 60 million years ago, moths and other nocturnal insects owned the night sky, flitting about unmolested by predators. The appearance of bats forced them to evolve a novel antibat strategy—a way of hearing the echolocating calls of hunting bats, in effect a radar detector. Moths with this strategic advantage are 40 percent less likely to be eaten, Dr. Fullard said. When they hear a bat speed up its sonar beam, indicating that it has begun its final approach, the moths launch into a wild acrobatic act, looping up, down and around to evade the less agile bat. If that is not enough, the moths fold their wings and dive to the ground, where the bat is hard put to find them.

Other moths emit their own sonar signal, blasting out high-pitched clicks that startle the bat. Still others use an auditory guise to dissuade bats by mimicking a sound produced by a foul-tasting moth.

But some tropical bats are like Stealth fighters, acoustically invisible to moths. They emit echolocation calls "so high-pitched that even the sensitive ears of moths cannot detect them," Dr. Fullard said.

Dr. Paul A. Faure, a biologist at Cornell University, has studied these "whispering" bats, which forage among vegetation rather than in the air. By implanting electrodes into the auditory nerves of the moths, he showed that the insects were deaf to the bats' high-pitched emissions.

Others hunt without making any noise at all. Dr. Z. M. Fusessery of the University of Wyoming reported that pallid bats are silent hunters. The enormous ears and finely tuned auditory systems of these bats are so sensitive they can hear an insect as it walks and detect the wing flutter of a moth about to take off.

Frog-eating bats have evolved yet another strategy, Dr. Tuttle found. They simply tune in to the frog's mating call and silently pick off the amorous amphibian.

The extraordinary vulnerability of animals that congregate by the millions in a few well-defined locations demands fast action from those trying to spare the woodsman's ax. Kirk W. Navo of the Colorado Division of Wildlife, who reported that "about 2,000 of the mines in Colorado have already been sealed without any input from wildlife experts," said he and his colleagues have recruited and trained dozens of volunteers to survey mine sites for bat life. Armed with bat detectors that pick up the animals' sonar, the volunteers work night shifts to monitor bat activity before the mines are closed. But with more than 20,000 inactive mines in the state, the volunteers are hard put to keep up with closings.

Two years ago, Dr. Tuttle said, one of the largest remaining bat populations in New Jersey, a hibernating hoard of 20,000 little brown bats, was sealed into a rural mine. A member of Bat Conservation International noticed the closing and called headquarters, which got the mine reopened in a few days. The mine and its surroundings have since been made a protected reserve for bats.

"Bat gates," vandal-proof metal grids that allow access to bats but not people, are increasingly being used to protect caves and mines that are major bat roosts and hibernation sites. Dr. Tuttle explained that each time human visitors arouse a hibernating bat, it uses up 10 to 30 days' food supply. More than a few arousals per winter can easily result in an entire colony starving to death before insects reappear in spring.

Dr. Elizabeth D. Pierson, a bat researcher from Berkeley, California, reported the successful relocation of a bat colony by the Homestake Mining Company, which was planning to reactivate an area containing a bat-occupied mine. After the colony emerged at night, the mine shaft was sealed and the colony moved into another mine nearby that was protected by a bat gate.

Finally, Tracey Tarlton of Bat Conservation International has been analyzing bridges to uncover design features that make ideal bat roosts. Instead of killing off bridge-inhabiting bat colonies "because bridge workers are afraid of them," she proposed structural features for bridges and overpasses that could provide safe haven for these valuable animals.

—JANE E. BRODY, October 1991

Amid Maze of Signals, Bats' Brains Can Form Precise Images of Prey

PEOPLE WHO GREW UP before television had little difficulty "seeing" the Green Hornet, the Lone Ranger or Baby Snooks on radio. But can insect-eating bats, relying solely on sound signals to find food, form comparable mental images of their prey to help them distinguish a delicious moth, say, from the leaves it flutters through?

Neurobiologists at Brown University have found that a bat can in fact form quite precise acoustical images of the many objects in its aerial environment. In the so-called tonotopic area of the auditory cortex, the part of the brain that processes sound signals, the scientists have pinpointed nerve cells that enable the bat to discern the exact size, shape and movements of insects in its path.

The probable role of these cells, called delay-tuned neurons, was deduced from experiments described in the journal *Nature* by Steven P. Dear, James A. Simmons and Jonathan Fritz of Brown's departments of neuroscience and psychology. They have been studying how sound is processed by individual cells in the brains of big brown bats, night-flying mammals a mere two and a half inches long that commonly cause panic reactions when they fly into people's attics and homes.

In an interview, Dr. Dear, a self-proclaimed bat lover, explained that from observing bats' behavior he came to suspect they were somehow forming mental pictures of their environments, the images being based not on sight but on sound.

Bats emit high-pitched squeaks, using the returning echoes to navigate and hunt. Researchers have shown, for example, that when bats first detect an insect, their sound emissions become shorter and more frequent, helping them zero in on their target. Somehow, these aerial acrobats seem to know the difference between an insect and other objects of similar size or density, suggesting that they can "see" objects by processing sound signals.

In laboratory studies done decades ago by Donald Griffin and colleagues, then at Harvard University, bats were first trained to catch mealworms tossed in the air. Then the researchers tossed the worms in amid a handful of Necco wafers, candy disks that throw back echoes similar to those from the mealworms. Yet the bats ignored the wafers and headed straight for the worms.

Other experiments have shown that a bat can fly easily through strands of very fine wire strung across its path, without having to move its head back and forth to check out the echoes from these obstacles. Dr. Dear concluded that "the bat seems to form a clear picture of what's out there when it is as much as a meter away from the wires."

"Furthermore," Dr. Dear said, "if bat behavior is recorded in the wild, bats can be seen flying after prey while continually aiming their heads directly at insects even though many other things are throwing back echoes." Thus the scientist deduced that the bat must have some way to know precisely what it is hearing, a mechanism that would allow it to "see" the insect.

He explained that "if echolocating insectivorous bats hunted only in wide-open spaces, there might be no need for such a mechanism. But bats often hunt in very cluttered environments, amid tree branches, for example, that create lots of echoes." Yet the hunting bat is able to distinguish between the branches and an insect destined to become its prey. Bats also have to navigate in narrow spaces to enter their roosts, a task that would be greatly eased by an ability to form an image of the cramped environment.

The bat's brain is replete with neurons that are tuned to process the time that elapses between the emission of a sound by the animal and the return of the echo after that sound hits an object. Using these "delay-tuned" neurons, the bat calculates the distance of the object that sent back the echo. Different neurons in the tonotopic region of the bat's auditory cortex are tuned to receive signals with different delay times, which means that some neurons respond only to echoes that return with long delays, indicating that they bounced off objects at a distance. Other neurons respond only to echoes with short delays that bounced off an object close by.

These short-delay neurons then shut down, but as the bat continues flying, new short-delay neurons fire in response to objects that are now close by. The bat can thus "see" at one time what is near, what is far and what is still farther away, forming a complete picture of its environment.

But the talents of the tonotopic neurons do not stop there, the Brown studies showed. The researchers found that the later in time that these neurons fire, the more selective they are in responding to a delay between the emitted sonar and the echo. This increasing selectivity indicates that the bat's brain acts as a fine-tuning device that helps the animal to refine the image of the objects it is approaching.

"This fine-tuning ability helps the bat determine how objects it perceives are related to one other," Dr. Dear said. "The mechanism is identical to a process used in computer vision to analyze images."

"Of course," the neurobiologist conceded, "we have no way to prove that the bat forms mental images." Unlike a person blind from birth, a bat cannot be asked to describe or draw pictures of objects learned through nonvisual senses. But, the researcher added, "at least we now know the bat has a way of encoding a scene, which may be the basis of memory."

—JANE E. BRODY, August 1993

Some Male Bats May Double as Wet Nurses

Scientists casting their nets in the dense forest canopies of Malaysia have discovered the first example of a wild male mammal that lactates. The species is a Dayak fruit bat, a large and poorly understood creature with an 18-inch wingspan, a doglike face and, it turns out, a touch of androgyny.

When the researchers captured a group of the bats in a wide-ranging effort to survey animals that inhabit the Malaysian canopy, they were dumbfounded to see that the eight adult male Dayaks in the net all had visibly swollen breasts that produced milk upon being gently squeezed.

"They looked like perfectly good males with large testes, but from the other end I could see they also had well-developed breasts," said Dr. Charles M. Francis, a research associate with the Wildlife Conservation Society who helped write a report on the bats that appeared in the journal *Nature.*

No other wild male mammals have ever been reported to lactate, the scientists said, although inbred domesticated male goats and sheep, in rare cases, have been found to make small quantities of milk, possibly because of a genetic mutation. But the Dayak fruit bats, whose Latin name is *Dyacopterus spadiceus,* are free ranging and seem perfectly healthy.

"What this paper describes is that the males have the plumbing and the physiological capability to lactate," said Dr. Thomas H. Kunz, a biologist at Boston University who is the senior investigator on the report. Though the amount of milk detected was only a tenth of the amount produced by females. "We haven't seen the males nursing, and we don't know anything about their social and mating systems."

He and other scientists said the males could be producing milk as a result of eating phytoestrogens, estrogenlike compounds found in some plant leaves. Dayak bats are thought to subsist mainly on fruit, but their diet could include leaves rich with estrogens; in theory the estrogens could stimulate milk production. In that case, the male milk could be a byproduct of the bats' diet, and serves no functional purpose.

Alternatively, the bats could have been exposed to pesticides or plastic residues that may mimic the effects of estrogens on the body.

But the researchers doubt that such pollutants account for the appearance of milk-rich breasts among male Dayaks. For one thing, they said, the two forest areas in which the bats were found are reasonably unspoiled; for another, estrogen mimics have mostly been blamed for reproductive defects, and the testes of the adult male Dayaks were found to be full of normal sperm.

The researchers said that the only circumstance under which male bats would be likely to nurse was if male and female Dayaks were monogamous and contributed jointly to the rearing of their young.

That sort of mating system accounts for less than 3 percent of all species. More often, male mammals spread their seed as widely as possibly. Only two bat species, the false vampire bat and the yellow-winged bat, are thought to form monogamous pairs.

But among bat lovers, the idea of a wet-nursing father is not unthinkable. Their nearly 1,000 species fill almost every ecological niche imaginable.

"Bats are extremely diverse and sophisticated," said Dr. Merlin D. Tuttle, an ecologist and founder of Bat Conservation International in Austin, Texas. "They've surprised us so many times before that I've learned to expect the impossible."

—Natalie Angier, February 1994

Pronghorn's Speed May Be Legacy of Past Predators

RACING AT TOP SPEED across the western plains, as close to flying as four hooves can take it, an American pronghorn in motion is a biological marvel, running nearly 60 miles an hour, faster than anything else on the continent.

Pantheon of Ghosts
At various times, three extinct species, the giant short-faced bear, *Arctodus simus;* the long-legged hyena, *Chasmaporthetes ossifragus;* and the North American cheetah, *Miracinonyx trumani,* swiftly pursued both the North American pronghorn and its direct ancestors.

Michael Rothman

A quick dash and the antelope easily shakes off even the most determined coyotes and wolves, presenting biologists with a high-velocity mystery. What is this perfection of running speed doing here where there is no creature capable of pursuing the chase?

After studying pronghorns for 14 years on the National Bison Range in Montana, one researcher says he has the answer. The scientist, Dr. John A. Byers of the University of Idaho in Moscow, says the pronghorn runs as fast as it does because it is being chased by ghosts—the ghosts of predators past.

In a book published by the University of Chicago Press, *American Pronghorn: Social Adaptations and the Ghosts of Predators Past,* Dr. Byers argues that the pronghorn evolved its heady running prowess more than 10,000 years ago, when North America was rife with fast-cruising killers like cheetahs and roving packs of long-legged hyenas.

"The realization just grew and grew that I was looking at an animal that was adapted to this former world," Dr. Byers said. "These were predators that would have been really, really nasty."

As researchers begin to look, such ghosts appear to be ever more in evidence, with studies of other species showing that even when predators have been gone for hundreds of thousands of years, their prey may not have forgotten them.

"It's going to be a very controversial idea," Dr. Richard Coss, a behavioral ecologist at the University of California at Davis, said of what he calls relict behaviors. Researchers used to thinking of behavior as infinitely adaptable and very quickly evolving "may not find the idea of relict behaviors comfortable," he said.

Though controversial, the idea is far from new. As seems to be true of every interesting notion in evolutionary biology, Charles Darwin explored this possibility himself more than 100 years ago. Darwin speculated on whether behaviors suited for life in the wild might persist for long periods in domesticated animals no longer subjected to the natural rigors faced by their ancestors.

Until 10,000 years ago, the rigors of life for pronghorns appear to have been extreme.

In order to survive, at one time or another these doe-eyed cud-chewing creatures had to evade the North American cheetah, the giant short-faced bear, a long-legged creature that was probably an impressive runner, as well as lions and jaguars, which were even bigger and faster than they are today. The young and the weak faced an even greater array of dangers with saber-toothed cats roaming about as well as numerous types of wolves and plundering dogs.

But the worst, by far, were the hyenas equipped with cheetahlike limbs and huge jaws full of ripping teeth.

"I don't think there's a predator alive today," said Dr. Byers, "that would've been as ferocious as that long-legged hyena would've been—its back as high as a person's waist and running in packs as it probably did. It would have been able to get on a group of pronghorns and drag them down and rip them apart. It would have been truly formidable."

Once the pronghorn is envisioned amid such predators, its speed seems much less extraordinary and much more obligatory, as it is hard to imagine any save the fleetest getting out of the Pleistocene alive.

And fast they became. One researcher clocked pronghorns at 55 miles an hour, though biologists guess that they may go even faster. Others tracking pronghorns by light plane going at 45 miles an hour say they have seen pronghorns put on a little burst of speed and effortlessly slip away. Pronghorns are endurance runners as well, going at 45 miles an hour for several miles without showing any signs of exhaustion.

Despite the explanatory power of ghosts of predators past, those studying behavior rarely stop to consider such vanished enemies, which was one of the major reasons Dr. Byers said he decided to write the book.

Dr. Daniel Rubenstein, a behavioral ecologist at Princeton University, called Dr. Byers's work "high quality" and agreed, saying that typically behavioral researchers would only consider such historical explanations if forced to do so. "Digging back into the past to see why traits originated is almost a last resort," said Dr. Rubenstein.

Part of the reluctance to entertain such ideas is the difficulty of testing hypotheses in which the principal players are all extinct.

Dr. John Fryxell, a behavioral ecologist at the University of Guelph in Canada, who called the book "convincing," said historical explanations were always more difficult to test. Just as scientists cannot repeat and manipulate the "big bang," researchers theorizing about long-dead predators cannot remove and rearrange extinct hyenas. They must instead examine whether other modern-day evidence supports their theory of the predators' importance.

In Dr. Byers's case, there is much supporting evidence, as many aspects of pronghorns' lives appear to have been shaped by their ancient enemies.

Many animals that rely on grazing will herd, roaming about in large groups. A group affords more eyes to spot an approaching predator and for any one individual in the group the chance of being attacked is diluted. So herding animals suffer the inconveniences of crowding, including struggling for food and dominance, in exchange for the safety of numbers.

Yet though adult pronghorns have nothing to fear from the carnivores around them now and nothing to gain from herding, they continue to do so in what appears to be another adaptation to ancient threats.

Dr. Byers says he has found hints of the past in pronghorn mating as well. If peak speed and endurance were once key to survival, one might expect pronghorn females to pick fathers for their offspring on the basis of vigor and athletic prowess, which is exactly what females do. When the mating season begins, males work to herd groups of females.

But as soon as a female is ready to mate she begins to attempt escape, leading the male on sprinting chases and drawing the attention of challenging males. Females typically stand back and watch as males struggle to fend off challengers, then choose the victor as their mate.

Today the pressures on the pronghorn population are predation on fawns by coyotes, bobcats and even golden eagles, as well as hunting of adults by humans.

It is not only pronghorns that hang on to ancient defenses. Dr. Coss and his colleagues found that California ground squirrels from populations that have been free from snakes for 70,000 to 300,000 years still clearly recognize rattlesnakes. Exhibiting stereotypic antirattlesnake behavior, the ground squirrels approach with caution, throw dirt and fluff up their tails. But fear, even of snakes, does not last forever. Arctic ground squirrels in Alaska, free of snakes for some 3 million years, seem unable to recognize the threat of a rattlesnake. These hapless squirrels exhibited only a disorganized caution, even after being bitten repeatedly.

Dr. Susan A. Foster, an evolutionary biologist at Clark University in Worcester, Massachusetts, and her colleagues have also found relict behaviors in stickleback fish. Working with a population that has long been free of sculpin, a dangerous predator, researchers presented preserved sculpin to the fish. To their surprise, researchers saw the sticklebacks immediately engage in stereotypic antisculpin behavior, treating the predatory fish with caution, avoiding its mouth and swimming behind to bite it.

In what will surely be the most controversial of the new studies, Dr. Coss and his colleagues are searching for relict behaviors in humans.

Researchers questioned three- and four-year-olds and adults about childhood nighttime fears. While the overwhelming majority of males reported being fearful of attack from the side, the greatest number of females reported being fearful of attack from below. Dr. Coss says these differences may be due to the life patterns of ancient hominid ancestors.

According to some theories, early female hominids were more adept climbers (evidenced today, in part, by the greater flexibility of the young adult female foot) and spent more time in trees than males. More likely to sleep in elevated roosts, females were most vulnerable to attack from below. Males sleeping on the ground, however, would have been more vulnerable to nighttime attack from the side.

Some might suggest that researchers were most likely detecting the ghost of television and movies watched. But Dr. Coss, while acknowledging the powerful effects of these media, suggests the opposite. Hollywood, he says, may be capitalizing on the primal fears that humans still carry from the days when they were easy, tender targets for many a predator.

—CAROL KAESUK YOON, December 1996

Tracking Down Michigan Moose No. 9 for a Six-Hundred-Mile Checkup

HIS EARS, YELLOWISH BROWN, were as soft as velvet, all 10 inches of them. His staring eyes were huge. He let out a low snort as his broad nose was stroked. His long legs were folded under his enormous body in the deep snow. This was Michigan's Moose No. 9 in repose, briefly tranquilized by darts carrying a powerful narcotic.

It had taken wildlife biologists on snowshoes upward of two hours to track him, using handheld radio directional antennas tuned to a transmitter attached to a leather collar on a second bull moose, who was quite probably one of No. 9's offspring. The two had bedded side by side in a forest of balsam, black spruce, hemlock, silver maple and yellow birch in temperatures of zero and below.

Somewhere nearby lay their huge palmate antlers, spanning almost six feet. An aerial spotter had seen the moose with antlers three weeks earlier, but they had since shed them.

Earlier in the day, Dr. Steve Schmitt, a wildlife veterinarian who also holds a graduate degree in biology, sighted the uncollared bull on the move and shot him with a dart carrying three milligrams of carfentanil, a narcotic 7,000 times stronger than morphine. He was part of a group who was darting moose, preferably yearlings, to place radio collars on them, to monitor their movements, mortality and reproduction.

At first, the drug only slowed No. 9. Even so, he was moving three or four times as fast as the three biologists on snowshoes who followed. With their long legs and eight-inch-wide hooves, moose move easily in three feet of snow.

Droplets of blood in the snow distinguished his track from that of his companion, trotting a mile on parallel courses up and down gentle slopes to a stand of hardwood. Then he stopped and sank into the snow. Dr. Schmitt and a colleague fired two more drug darts into his side and rear, but when they approached, he staggered to his feet and charged.

"Get behind a tree!" a biologist shouted to a trailing reporter. But the bull veered and, after another 30 yards, sank down, finally stilled by the narcotic.

Moose No. 9 is something of a phenomenon. He started life a dozen years ago in the Algonquin Provincial Park in Ontario, more than 600 miles to the northeast. What brought him here was an unusual experiment in wildlife management: Operation Moose Lift. In 1985 and 1987, it transferred 59 moose from Ontario to the Upper Peninsula.

The project has proved a great success. The original 59 have increased to 265 and the survival rate of calves after the first year has been above 80 percent. More than 85 percent of the cows are pregnant from the autumn mating season, or rut. "We're very excited," said John Hendrickson, the regional wildlife supervisor of the State Department of Natural Resources, based in Marquettem on Lake Superior. "They're doing well. Public acceptance is one of the most gratifying things about it."

Moose No. 9 is one of about 51,000 of the huge herbivores inhabiting 17 states other than Alaska, which remains something of a moose paradise with upward of 144,000 of the species *Alces alces*. In the Pleistocene age, moose ranged as far south as the Carolinas, fossil finds have shown, but now they are restricted to the northern and mountain states. The latest moose census was prepared by Albert W. Franzmann of the Alaska Moose Research Center. His figures indicate that the population has more than quadrupled in the 17 lower states since 1948.

The idea of translocating moose, specifically to Michigan's Upper Peninsula, had been around for a long time. Indeed, there was an abortive attempt in the mid-1930s, which failed because of the poor condition of the moose and subsequent poaching. Except for the odd migrant from Canada, moose had been absent from the Upper Peninsula and Lower Michigan for almost a century, wiped out mainly by loggers and miners who killed them for meat.

The godfather of the latter-day translocation was Ralph Bailey, a regional wildlife biologist. In the early 1970s, he noticed that the Upper Peninsula's vast, heavily logged timberlands had grown back into forests that made good habitat for moose. That, with a notable decline in the deer population, convinced Mr. Bailey that "the time was right to reintroduce the moose to the Upper Peninsula."

"After all, moose were the original inhabitants here," he said. "There were very few deer before white men came." His lobbying, described by a colleague as "gentle persuasion," eventually succeeded.

Planned as precisely as a commando raid, Moose Lift, a joint operation of the Department of Natural Resources and Canada's Ministry of Natural Resources, started with a helicopter scaring moose out of the Algonquin forest onto a frozen lake. There they were darted by an airborne shooter. On the lake ice, a team wrapped each animal in a large nylon sling for transport by helicopter a dozen miles to specially made crates on waiting flatbed trucks.

The narcotic was then reversed almost instantaneously with an injection of a drug called Naltrexone. The drive to Michigan took almost 24 hours, after which the moose were uncrated. Some were released with radio collars, and some of the transmitters are still functioning. The total cost of the two lifts was about $150,000, much of which was contributed by sportsmen's clubs, wildlife organizations like Safari Club International and even a local Lions Club.

In gratitude, Michigan sent 100 wild turkeys to the Algonquin park, where there are now 700 birds and limited hunting of turkeys has started.

In the North Woods, the Department of Natural Resources, or DNR, is still mocked here and there as "Do Nothing Right" or "Damned Near Russia," but where moose are concerned, the department is viewed as a friend. Poachers—two have been apprehended—are deplored.

Moose consciousness is flourishing. A new summer camp is named Moose Hollow. Highway signs announce Moose Country, bumper stickers boast "Michigan: Where the Moose Run Loose." A dairy produces Moosetracks Ice Cream. People wear "Moose Booster" caps and sell shellacked moose droppings as "Moosletoes." The state has issued a brochure with a map showing where tourists might spot moose.

As a result of the monitoring in the forest west of Nestoria, Mr. Hendrickson said, Michigan biologists have "picked up problems of mortality that nobody knew about," including insufficiently developed fetal lung tissue. Fifty-one moose are known to have died, and the cause for 15 of them is a neurological disease caused by the parasite *Parelaphostrongylus tenuis,* or brain worm. Its host is white-tailed deer, whose feces are visited by snails, which carry the parasite in their mucus to vegetation that is moose fodder.

Tracking moose is strenuous work. A day on snowshoes in 30 inches of snow is like a day on a StairMaster, only colder. The temperature the first day was minus 20, with a wind-chill factor of minus 65. Still, Rob Aho, a wildlife biologist, was able to dart a male calf west of this tiny crossroads with his .32-gauge shotgun, whose projectile is propelled by a .22-caliber blank round. Four days and many arduous hours later, Dr. Schmitt darted Moose No. 9, and on the sixth day, Mr. Hendrickson shot a dart into a female calf near Tower Lake.

Each moose was collared, making a total of 19 in the range. Each received a topical antibiotic spritz on its dart wounds, a shot of long-acting tetracycline against infection and some ointment on its narcotic-glazed eyes. Then Dr. Schmitt injected the reversal drug and they were on their feet in a minute or so. "This is one of the high points of my career," Dr. Schmitt said.

Mr. Aho added, "It's what you dream about."

And Mr. Hendrickson said: "If there is another fourfold increase in the next 10 years, we might have 1,000 moose here. Then moose hunting could be a real possibility."

—DAVID BINDER, March 1994

Patricia J. Wynne

PCBs in the Food Chain
Polar bears eat seals; seals in turn eat cod that eat larger plankton that eat smaller plankton that eat algae that absorb PCBs.
At each step, the concentration of PCBs increases, and they become very concentrated in the bears' fatty tissue.

For Arctic Data, Ask a Polar Bear

THE HELICOPTER bucks and sways. Dr. Malcolm A. Ramsay is riding behind the pilot over the wind-whipped taiga of northern Manitoba. He anchors his left arm in the open window and aims a rifle at the lumbering yellowish white shape below.

The rifle gives a faint pop and a four-inch long Telazol tranquilizer dart jabs the shoulder of the big, hungry polar bear. Within four minutes it is temporarily immobilized.

The pilot, Steve Miller, drops Dr. Ramsay off in a clearing near the fallen animal and returns to the Churchill Northern Studies Center, about 75 miles to the north.

He will return in a few hours to pick up Dr. Ramsay and whatever he has learned about the ecology of the taiga, a coniferous forest in northern climes, from a brief visit with one of its most prominent citizens.

Because of its peculiar biology, the polar bear acts as a blotter for the stresses on the Arctic environment, and scientists are studying it to learn about chemical pollution, global warming and other potential threats to the survival of this species as well as humans.

Wrapped in layers of liners, a cavernous parka, thermal boots, thermal socks and thermal gloves, Dr. Ramsay, an associate professor of vertebrate ecology at the University of Saskatchewan and one of Canada's leading experts on polar bears, gives the sedated beast sprawled on the snow a physical as thorough as any a company doctor gives a corporate executive. With a winch, tripod, net and electronic scale, he weighs the bear. Other instruments measure its electrical resistance to determine its lean body mass, as opposed to adipose tissue, or fat.

With pliers he extracts a vestigial premolar tooth that the bear does not use to determine the animal's age. The tooth builds up rings like the trunk of a tree.

From the femoral vein, he draws blood samples for biochemical tests. He takes a sample of fat tissue for later analysis of polychlorinated biphenyls (PCBs), industrial pollutants that make their way into the polar bear's niche and become concentrated in its fatty tissue. Researchers already know that the levels of PCBs in the bears are dangerous and could start having effects on reproduction.

Biologists are now trying to ascertain the danger level for PCB concentration in polar bears. For seals, Dr. Ramsay says, deleterious reproductive effects are evident in the PCB levels of 70 parts per million that have been recorded in the blubber of Baltic Sea seals. These animals have been practically wiped out over the last two decades.

The Baltic is a known dumping ground for materials containing PCBs. Probably reflecting this, adipose tissue of polar bears in the Svalbard Islands north of Norway has recently been found to have a concentration of 32 parts per million. Hudson Bay is not yet such a dumping ground, and the mean level of PCB concentration in the Churchill area bears was most recently recorded at 8 parts per million.

Every year between August and the end of November, Dr. Ramsay comes to Churchill to study the species, *Ursus maritimus*. Over the last dozen years he has similarly examined more than 1,000 of the species.

Churchill is a popular place for the study because it is easily accessible, by rail or plane, and because the bears congregate in the area just before Hudson Bay freezes. Then they move out on the ice for the ringed seals that are their basic diet. Until the ice forms, they do not eat or drink for periods often exceeding four months.

"Why do I like to work with polar bears?" the professor asked. "I guess it's because they live in a very simple ecosystem. They feed on a single species, live in a simple environment. It's easier to understand the ecological principles at work in a simple ecosystem than it is in the Tropics, or the timber region, where interactions are more complex."

The knowledge he is accumulating about this "simple" system could prove important not only for an aloof and remote species roaming the roof of the world but for humans.

During their long fast, even longer than that of other bears, polar bears meet their energy requirements from fat reserves. These are extensively depleted. Every day of fast they burn up one to one and a half kilograms of fat.

But while the fat is burned off, the protein mass of the bear remains constant, or declines only gradually, because polar bears are able to recycle proteins internally. Humans and most other mammals show a continuous loss of these compounds while fasting.

Proteins make up the structural parts of the body, like muscle, bone, connective tissue, hair, skin and fingernails, and are the enzymes that cause reactions, like digestion.

In all mammals, including humans, old protein gets broken down into urea. Kidneys filter urea out of the blood and into the urinary bladder, from which it is excreted from the body in urine. The old protein is replaced by protein taken in through food.

Should kidneys malfunction, the buildup of urea creates a life-threatening danger. In such conditions, humans usually have to go through dialysis, having their blood filtered in effect by an artificial kidney.

But Dr. Ramsay's examinations have shown that polar bears are blessed with some kind of biochemical pathways that avoid use of kidneys. Instead, the urea of the bears is converted back into amino acids, the building blocks of proteins.

Just how they do this is a mystery that Dr. Ramsay thinks may be solved in the near future. He is working on finding the answer in a joint project with Dr. Ralph A. Nelson, chairman of the medical school of the University of Illinois.

If they find a special enzyme in the bear that facilitates the conversion of urea wastes back into amino acids, bypassing the kidneys, and if it can be reproduced pharmacologically, it might help humans with kidney failure to avoid dialysis.

Some of Dr. Ramsay's bears get collars emitting radio signals that bounce off satellites and can be picked up by telemetry receivers in the helicopter. This permits him to follow reactions in the same animal over a span of weeks or months. In selected bears next season, he will inject urea molecules labeled with a special nitrogen isotope. He will then try to find the same bears again to track those molecules through the biochemical pathways.

But ecology looms large in his research. He is studying the buildup of contaminants like PCBs, which are now found in the Arctic food chain. PCBs, banned in North America and most of Western Europe but still produced in Eastern Europe, Asia and other less developed areas, are apparently borne by winds to the northern seas and absorbed by marine life.

Polar bears, at the top of the food chain, eat the seals that eat the cod that eat the larger plankton that eat the smaller plankton that eat the algae that absorb the PCBs. At each rung of the food chain, the concentration of PCBs increases.

Once ingested, the PCBs become concentrated in the fatty tissue of the bears. "We don't yet know what happens to the PCBs when the bears fast and burn up their fat reserves," Dr. Ramsay said. "Are they broken down by the liver and excreted from the body? Are they broken down somewhere else? Are they concentrated in the remaining fat? Are they transferred to the cubs through the mothers' milk?"

He and graduate students who join him on some of his northern expeditions, like Susan Polischuk, a 25-year-old master's candidate, and Stephen Atkinson, 24, a Ph.D. candidate, are trying to find answers, joined by Dr. Ross J. Norstrom of Environment Canada, a leading expert on PCBs in food chains, who works closely with Dr. Ramsay.

Polar bears are not technically an endangered species. They are protected by law in all Arctic countries. In Canada, Indians and Eskimos are permitted to kill a certain number of the bears each year. Others may kill them only in self-defense.

Still, Dr. Ramsay worries about their susceptibility to "catastrophic events" in the future. The PCBs are only one danger. Global warming is another.

Any rise in the mean annual temperature changing the dynamics of sea ice could further shorten the time of their feed with disastrous consequences.

Another possible source of trouble is that their food supply might die off, as occurred recently in the North Sea when masses of gray and harbor seals perished from an epidemic of a morbillivirus, similar to the distemper that strikes dogs. In the Mediterranean thousands of seals as well as dolphins recently succumbed to a similar epidemic.

Still another threat is polar bears' limited genetic variability. This has been determined in studies of their genetic blueprint, DNA. Thus with any sudden change in their environment, they might not have the flexibility to cope.

—CLYDE H. FARNSWORTH, November 1992

Experts Race to Save Dwindling Rhinos

FOR TENS OF MILLIONS of years, the rhinoceros has rumbled across the world's landscapes. Squat, muscular, thick-legged and broad-chested, it is the very embodiment of raw animal power.

One species, a prehistoric behemoth that stood 18 feet at the shoulder, ranks as the biggest land mammal ever to live on Earth. And as recently as the last century, millions of rhinos still roamed Asia and Africa. But now, despite strong efforts by conservationists over the last few years, fewer than 11,000 of these powerful animals remain on the planet—the total is down by more than 80 percent in the last two decades—and the number continues to drop.

Some scientists fear that none of the five remaining species of rhino is large enough to maintain long-term genetic health. A desperate attempt to reverse the decline is under way, and some experts say the rhino's unfettered days as a creature of the wild may be numbered.

They are being crowded out of their natural habitats by development and slaughtered for their horns. These trophies fetch thousands of dollars in much of Asia, where they are prized as the ingredient of folk medicines, and in Yemen, where they are made into dagger handles.

Consequently, rhinos are now far outnumbered by the American bison, a species that for the most part exists only in protected enclaves. And like the bison, some experts say, the rhinoceros has reached a point where the only way to rescue it from extinction is to confine it to well-guarded, intensively managed sanctuaries where it can no longer roam over its natural range. "Mega-zoos," one expert calls these sanctuaries, which are being established in Africa and Asia. Only there, some conservationists argue, can rhino populations be manipulated to preserve not only their numbers but also the gene pools that will enable them to survive in the long run.

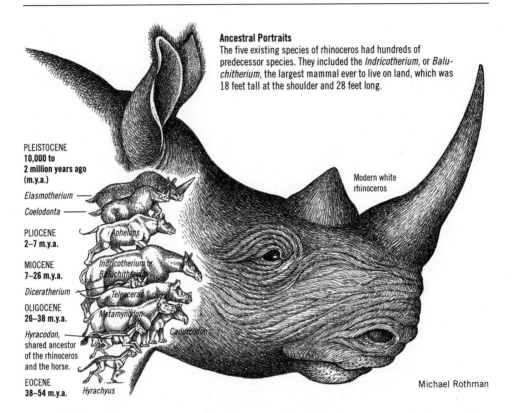

Ancestral Portraits
The five existing species of rhinoceros had hundreds of predecessor species. They included the *Indricotherium*, or *Baluchitherium*, the largest mammal ever to live on land, which was 18 feet tall at the shoulder and 28 feet long.

PLEISTOCENE
10,000 to
2 million years ago
(m.y.a.)

Elasmotherium —

Coelodonta —

PLIOCENE
2–7 m.y.a.

Aphelops

MIOCENE
7–26 m.y.a.

Indricotherium or Baluchitherium

Diceratherium —

Teleoceras

OLIGOCENE
26–38 m.y.a.

Metamynodon

Caenopus

Hyracodon,
shared ancestor
of the rhinoceros
and the horse.

EOCENE
38–54 m.y.a.   *Hyrachyus*

Modern white
rhinoceros

Michael Rothman

The growing effort to establish these sanctuaries is seen as a last resort now that other attempts to cut the rhino's losses to poachers and habitat destruction have fallen short. The goal is to reintroduce the rhinos to a fully free existence someday. But for now this is considered an uncertain prospect, given escalating human disruption of the landscape.

Is the rhinoceros about to disappear from the wild?

"I hope not," said Dr. Oliver Ryder, an expert on the rhino at the Zoological Society of San Diego. "But if we don't take this step, we're going to see the end of the rhino. Full stop." Dr. Ryder, a geneticist at the zoo's Center for the Reproduction of Endangered Species, was the organizer of a conference in San Diego, at which scientists and conservationists from around the world gathered to assess the status and future of the rhinoceros.

The outlook is not hopeless. In Asia, vigorous protection measures have enabled the great one-horned, or Indian, rhinoceros to recover slowly from near extinction early in this century to a population of about 2,000 today. Recent studies have found that the gene pool of this species remains healthy, and it is being successfully reintroduced to parts of Nepal, from which it had disappeared. Similarly, the white rhinoceros of Africa has been painstakingly brought back from the brink of oblivion. About 4,500 now exist, and it, too, is being reintroduced to its former range.

But poaching and habitat loss continue to chip away at populations of the Indian rhino, according to the World Conservation Union, and a political insurgency in the Indian state of Assam, where most of the Indian rhinos live, has clouded the animal's future. Two years ago, insurgents invaded a major sanctuary, killing guards and wresting the sanctuary from government control.

"The political instability on the Indian subcontinent doesn't provide a very good long-term prognosis" for the rhinos, said Dr. Thomas Foose, an expert on the breed who heads the captive breeding specialist group of the World Conservation Union. The same is true in Nepal, he said, where a democracy movement confronts the monarchy, and also in South Africa, where much of the white rhino population lives.

Political instability is just one of a number of possible threats, including disease epidemics and natural disaster, that could wreak havoc on populations that even in the case of the white rhino remain small, Dr. Foose said. A catastrophe of any kind could mean a giant, possibly irreversible, step toward extinction.

The other three rhino species, the Javan, the Sumatran and the African black rhino, are already staring extinction in the face. Scientists place the population of black rhinos at 3,000 to 3,500, down from about 3,800 in 1987; of the Sumatran, or hairy, rhinoceros, at about 700; and of the Javan rhino—the rarest large mammal in the world—at fewer than 60.

For these three species "it is the last stand," said Dr. Eric Dinerstein, a rhinoceros specialist at the World Wildlife Fund.

And for rhinos globally, said Dr. Foose, "the general trend is downward; right now it's a very desperate situation."

All five species have been declared endangered under an international convention barring trade in rhino horns.

The rhinoceros dates back 60 million years. Starting out as a small, three-toed relative of the early horse and looking much the same, it evolved into hundreds of species and forms. It first appeared in North America and made its way eventually to Asia and Africa. It reached something of a zenith about 35 million years ago with the appearance of the *Indricotherium,* also called the *Baluchitherium,* largest of all mammals. This hornless rhino weighed nearly five times as much as the biggest known elephant and could browse among tree branches 25 feet above the ground.

Its modern descendants, the largest of which are about a third as tall as the *Indricotherium,* are second in size only to the elephant among today's land animals. Its armored appearance brings to mind the image of a tank; its irresistible charge, the image of a locomotive. It can run as fast as a horse, about 35 miles an hour, and like the horse, the rhino is a vegetarian.

It is solid muscle.

The distinctive horn of the modern rhino is made up of thousands of tiny strands of keratin, the substance of which human fingernails are composed. Rhinos can grow a second horn if one is lost, and they use the horn to plow up ground while foraging and as a weapon.

They live relatively long lives—about 40 years in the case of the African rhinos—which means that any genetic deterioration would take place relatively slowly. That is a conservation plus. But they have relatively few offspring—one every two or three years for the African rhinos, at most—which means that it takes a long time to rebuild populations once they are depleted.

In Asia, the population of Indian rhinos was sharply reduced by sport hunters in the 19th century. Overhunting, combined with habitat destruction caused by agricultural expansion, originally brought the Indian, Javan and Sumatran rhinos low, and in recent years both habitat destruction and poaching have maintained the pressure.

Rhino hunting in Africa, like elephant hunting, was strictly licensed and controlled by colonial administrators, but rhinos were extensively eradicated to make way for human settlement and farms. Estimated conservatively, there were 2 million to 3 million black rhinos in Africa at the turn of the century, said Dr. David Western, a Nairobi-based expert on both the rhino and elephant for Wildlife Conservation International, an arm of the New York Zoological Society.

Poaching in search of the valuable horn is the main cause of the black rhino's most recent decline. Twenty years ago, there were an estimated 60,000 to 70,000 black rhinos; as recently as 1980, there were 15,000. Today their numbers are so few, said Dr. Western, that the loss of only 400 to

500 animals a year would more than offset births and continue to drive the population down. Trading in rhino horns has been banned internationally since the 1970s, but it continues nonetheless. Unlike elephant ivory, Dr. Western said, rhino horns are very easy to smuggle, and the market is "extremely diffuse," making it hard to halt the continuing trade.

A worldwide ban on trading in elephant ivory has been in effect since late 1989, and it has been credited with largely shutting off the trade. That is an effective strategy, said Dr. Western, since about 600,000 African elephants remain. But with so few black rhinos, even a trickle of illegally traded horns can have a major impact on their population.

The result is that black rhino populations have declined by 98 percent in parts of East Africa in the last 15 years. Efforts to combat poachers throughout the black rhino's natural range have proved frustrating, said Dr. Ryder, because the available antipoaching forces are spread so thinly over the animal's vast territory.

The emerging new strategy, both in Africa and Asia, is to sequester the rhinos in small, defensible sanctuaries. A similar strategy has enabled the North American bison to rebound from a few hundred in the late 19th century to an estimated 75,000 today. Black rhinos are being moved into sanctuaries throughout Africa, and there are already signs that where the approach has been tried, it is having an effect.

How long will this kind of management be required? As long as human population growth and development continue to disrupt the ecosystems where rhinos live, some conservationists say.

It may still be possible for rhinos sequestered in sanctuaries to live somewhat as they would in nature if humans had never entered the picture. "But as far as being far-ranging and completely unfettered by man," said Dr. Ryder, "that's completely unrealistic at this time."

—WILLIAM K. STEVENS, May 1991

Even Shorn of Horns, Rhinos of Zimbabwe Face Poacher Calamity

THE BLACK RHINOCEROS wore a radio collar and her horns had been shorn with a chain saw to make her less valuable.

Even so, Million Sibanda shouldered an AK-47 assault rifle as he circled in, trickling dust through his fingers to make sure he was downwind. The gun was not for the rhino, but for poachers, who would kill the beast just for the pathetic stump remaining on her face—and would kill a park scout like Mr. Sibanda for being in the way.

Mr. Sibanda and Stewart Towindo, a park ecologist, crouched 40 yards from where the rhino and her bull calf stood browsing, like two gray frigates moored among the thorn bushes, and spoke in a whisper. "I'd rather have them with horns," said Mr. Towindo, gazing wistfully at the defaced animal, and shaking his head at what man has done to nature in the name of saving the rhino.

In the war for the future of the black rhinoceros, one of the planet's most ancient and endangered mammals, Zimbabwe has been an embarrassing rout.

Even the leaders of the conservation campaign use phrases like "spectacular failure" to describe the country's calamitous five-year decline from Africa's richest haven, with as many as 2,000 black rhinos, to a ravaged population of fewer than 300 today.

Beginning in May 1992, Zimbabwe darted every rhinoceros it could find with a tranquilizer gun and sawed off its horns, on the theory that poachers would bypass a hornless animal.

But so dramatically have Zimbabwe's tactics failed that the country now proposes a radical new approach: undercutting the poachers by legalizing trade in rhino horns, which are prized in Asia, where they are

46

ground into a fever-reducing potion and in Yemen for ceremonial dagger handles.

Mike Kock, the state veterinarian who oversees the rhinos in Zimbabwe, envisions state farms where herds of rhinos would be harvested like flocks of sheep. The horns grow back about three inches a year.

"This is one example of a ban that has failed completely, and it's failed because the demand is too great," he said. At a meeting in Fort Lauderdale, Florida, of the rhino committee of the Convention on International Trade in Endangered Species, which is the main international treaty on wildlife trade, South Africa and Zimbabwe plan to support a resolution lowering the protected status of the more plentiful white rhino, as a first step toward legal trade in the horns.

Despite the new respect South Africa commands after the election of President Nelson Mandela, the committee is virtually certain to reject the proposal. The public outcry would be too great, and even among wildlife officials in southern Africa, where commercial use of wildlife is a favored method of conservation, there is debate about whether it would drive the price low enough to put poachers out of business.

Critics say the enthusiasm of Zimbabwe and South Africa for legalization has much to do with the fact that the two countries are sitting on tons of rhino horns, stockpiles amassed from contraband and dehorning that would be worth millions of dollars if the ban were to be lifted.

In Zimbabwe's case, the critics say, it is not the ban that has failed but the government, which has been unwilling to do the one thing that does seem to save rhinos: spend money.

"The Zimbabwe budget for national parks in 1981 was eighteen million dollars," said Esmond Bradley Martin, a Nairobi-based consultant to the World Wildlife Fund, a conservation organization in Washington. "It is five million dollars today. Since the early 1980s, the government has continually put less and less into their parks. That's why it's been a disaster."

As a special United Nations envoy for rhinoceros conservation last year, Mr. Martin pressed Zimbabwe unsuccessfully to do what most other countries rich in wildlife have done: raise park admission fees, at least for affluent foreign visitors, and earmark the money for conservation.

By spending money on protection, Mr. Martin said, South Africa has increased the number of black rhinos from a few dozen to about 900, now the largest number in Africa. Namibia and Kenya, which invested heavily in intelligence networks to foil poachers, have also made headway.

Thanks to those countries, the black rhino's plummet toward extinction, from 65,000 in 1970, is thought to have leveled off at around 2,500 today.

But Zimbabwe has not given its parks the same priority. It charges visitors a fraction of the fees demanded in other countries (or at the private resorts in Zimbabwe itself). Admission to this park, for example, costs $2.50, and a cozy bungalow for two people rents for $15 a night.

Wildlife officials here agree that Zimbabwe has starved its parks, but say they have no incentive to lobby for higher fees. The money the parks generate goes to pay other government expenses, including a growing military budget.

Glenn Tatham, the chief warden of Zimbabwe's parks, says his staff is demoralized by low pay, danger (four rangers have been killed by poachers in the last decade) and the lack of success. They are up against seasoned killers, who perfected their skills by exterminating the rhinos of Zambia, then moved south across the Zambesi River in search of new hunting grounds.

Rangers here say the poachers may cross the long, poorly policed border, kill a rhinoceros, hack off the horn and disappear. Or they may camp and prey on a park for months before lugging their booty back to the well-established wholesale horn market in Lusaka, the Zambian capital.

Mr. Tatham says the fact that no rhinos have been poached here since February is a misleading consolation. "Yes, there's been a de-escalation of rhino poaching," he said. "But there are very few rhinos left to poach."

Although in theory the dehorning might have deterred poaching, no sooner was it complete then the department ran out of money to pay its scouts for four months. Poaching gangs swarmed into this, the largest national park, slaughtering most of Hwange's dehorned white rhinos and many of the hornless black ones.

Dehorning, Zimbabwe learned, did not make rhinos invulnerable. Poachers killed for the stump. They killed to save themselves wasted time. They killed because in the thick brush they could not tell whether a rhino had a horn.

Although there is no evidence for it, Mr. Tatham and others speculate that horn syndicates may have ordered dehorned rhinos killed to raise the value of hoarded horn. "When the last rhino gets shot, or dies from loneliness, the horn will be like diamonds," the chief warden said.

After watching Zimbabwe—and listening to the outcry from animal lovers scandalized by the buzz-saw surgery—other countries with rhinos, aside from Namibia and tiny Swaziland, have declined to take up dehorning. Some studies have also suggested that dehorned mothers are less able to defend infants against predators.

Although most rhino specialists still believe dehorning can help reduce poaching, they agree it is useless without policing and a costly, periodic trimming of the stumps.

"What went wrong was they didn't spend the money to look after the rhino after horns were removed," said Mr. Martin. "And you have to re-dehorn. For Zambian poachers a year or year and half of horn is worth the risk."

Zimbabwe's latest tactic has been to relocate the rhinos to smaller areas where they can be better guarded. Most of the country's black rhinos have been moved either to three preserves on private farms or to four public "intensive protection zones" like one here in Hwange, where the park service can concentrate its limited manpower.

"This is the last stronghold within the stronghold," said Mr. Tatham.

For easier guarding, all the rhinos who roam these zones are being fixed with collars holding transmitters that can be tracked with a directional antenna.

But Norman English, the warden who presides over the protection zone in Hwange, where perhaps 50 rhinos wander over 800 square miles, said the program is already falling short. Instead of the 67 scouts he expected to have patrolling his domain, he has 38. For this perilous work, a veteran scout earns about $50 a month.

"You can have all the technology you like," he said. "It's not going to save rhinos unless you've got guys on the ground."

—BILL KELLER, October 1994

2

THE REALM OF RODENTS

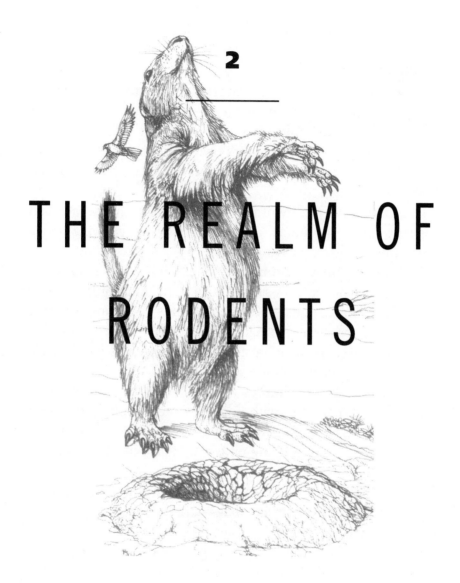

The quintessence of being a rodent is the stellar ability to gnaw. Their incisor teeth grow steadily throughout their lives. Unless worn down by constant gnawing on hard substances, like nuts, bark or telephone cables, the teeth will grow past each other and the animal will be unable to feed itself.

Another peculiarity of rodent dentistry is that the jaws can operate in two positions, a forward mode for gnawing and a backward mode for chewing.

This trick may seem nothing much, but it has enabled rodents to become the most successful of all mammal orders. About half of all living mammal species are rodents, including squirrels, chipmunks and marmots; field mice, deer mice, voles, lemmings and muskrats; rats; mole rats; chinchillas; coypus; guinea pigs; capybaras; porcupines; beavers and dormice.

Some rodents are impressive architects. Many make elaborate burrows, complete with nests, storage chambers and back exits. The beaver's dams, built from gnawed-down trees and waterproofed with mud, are on a scale that can alter the landscape.

Rodents have experimented with gigantism—the capybara of South America is four feet long—and with sociality. Mole rat societies have a queen and workers, somewhat like the arrangements of ants. But the distinguishing feature of the order is perhaps its very lack of specialism. Rodents tend to be adaptable generalists.

One of the tickets to success in the modern world is to live with humans, or at least in the environments humans create around themselves. Rats and mice have scored a remarkable victory, in ecological terms, in learning how to thrive in the crannies of human habitations. They have colonized not only the farm and the barnyard, but intensely urban environments such as the sewers and subways.

By the criterion of population numbers, certainly one of evolution's yardsticks, rats are as successful as humans: The rat citizenry of the United States is estimated to number as many members as its human counterpart.

———————————

Female Gerbil Born with Males
Is Found to Be Begetter of Sons

FOR MANY A YOUNG GIRL, a childhood spent with brothers has moments of fear and exasperation. Yet the most overbearing human brothers cannot begin to compare with Mongolian gerbils, creatures that can utterly change the fate of their sisters before any of them have even been born.

Scientists studying the blond, fuzzy-tailed rodents have discovered that when a female fetus matures in the womb with a male fetus on either side of her, the impact of all those male hormones on her development has a startling consequence: The female grows up and bears litters with a significantly higher proportion of sons than do females who spend their prenatal days sandwiched between two sisters.

Nor does the influence of brotherly hormones end with the female's immediate offspring. In bearing an excessive number of males with each pregnancy, the mother also helps ensure that any daughter she carries is herself likely to be surrounded by brothers, and hence to be exposed to the high levels of male hormones, or androgens, that will turn her into a vigorous begetter of sons. Gerbils generally have seven to eight pups per litter, and the study showed that those females whose immediate womb mates were brothers end up producing broods that are about 60 percent male.

The impact of siblings on the rodents' offspring can also work the other way. Female gerbils positioned in the womb between two other females develop in an environment that is especially rich in female hormones, or estrogens, and as a result they end up bearing slightly more daughters than sons. Those daughters in turn are prone to give birth to extra females.

The work, appearing in the journal *Nature*, overturns widespread scientific assumptions that inherited traits are invariably relayed through the genes.

"The most important lesson here is that just because you see a physical concordance between mothers and daughters, you can't automatically attribute it to genetic factors," said Dr. Mertice M. Clark, a psychologist at McMaster University in Ontario, Canada, the lead author of the report. "There could be some other mechanism at work." Dr. Clark performed her experiments with Dr. Bennett G. Galef, Jr., and Peter Karpluk, also of McMaster University.

Scientists have known for some time that the hormonal environment in which a fetus develops can influence its body, brain and behavior, but the latest study offers the strongest evidence that hormonal exposure can have an impact lasting many generations.

"This is a fascinating study, and a wonderful demonstration of how naturally occurring hormones have broad effects on fetal development," said Dr. John G. Vandenbergh, a zoologist at North Carolina State University in Raleigh, who wrote a commentary accompanying the report.

The discovery also gives scientists a fresh insight into how sex ratios are determined in mammals. For years, researchers have been trying to breed animals that spawn either mainly daughters or mainly sons, but to no avail. That may be because, in seeking a genetic trait associated with a particular sex ratio, the researchers were searching for something that does not exist. "There's been no successful attempt to select genetically for a sex ratio," said Dr. Clark. "It's possible that sex ratio is under hormonal control. It's possible that a high level of testosterone in the mother, or the father, is associated with sons," while high levels of female hormones are associated with an enhanced output of daughters.

The researchers do not yet understand how hormonal influences during fetal developmental end up affecting the sex ratio of the gerbils' offspring when they reach adulthood. Scientists propose that an excess of androgens or estrogens somehow influences the character of the female's eggs, changing the thickness or permeability of their membranes and making them more susceptible to penetration either by sperm bearing a Y chromosome—the hallmark of maleness—or an X chromosome. Alternatively, hormonal exposure in the womb may affect the character of the female's uterus, making it more hospitable to either male or female embryos.

And while the work has no immediate relevance to people, Dr. Clark and others pointed out that physicians are just beginning to wonder if, among humans, exposure in the womb to potent hormones may have an impact extending beyond a single generation. For example, the children of mothers in the 1950s and 1960s who took diethylstilbestrol, or DES—a synthetic estrogen—while pregnant have had a host of health problems, from an increased risk of rare cancers to infertility. As the children of DES children begin reaching reproductive age, it remains to be learned whether they may suffer any lingering effects from the medication their grandmothers had taken.

In experiments performed over the last 15 years on different species of rodents, Dr. Vandenbergh and many others have observed the startling impact of androgens on female fetuses. Male fetuses begin generating testosterone and related hormones early in development to aid in sculpturing and refining their own masculine forms, but if a female is very close to the male she, too, may be exposed to noticeable amounts of the androgens. Female rats squeezed between brothers end up with brains that are somewhat masculinized, particularly in the certain parts of the hypothalamus, a region known to differ between males and females of many species and now the site of a ferocious debate over the origins of homosexuality in humans.

The female rodents also end up with a more masculine style of behavior, roaming larger distances and marking a greater area as their territory than do most females. However, they mate happily with males and are perfectly competent mothers.

Scientists have made many attempts to understand sex ratios in animals. Among many amphibians, reptiles and fish, temperature is a key to gender. For example, in the Atlantic silverside fish, larvae that develop in cold water become females, while those that grow in hot water turn into males. Among mammals that bear only one baby at a time, Dr. Robert L. Trivers of the University of California at Santa Cruz and other scientists have suggested that the health of the mother is a factor in the sex of the child: The healthier the mother, the more likely she is to bear a son. Perhaps not surprisingly, this theory remains sharply contested by many biologists, and the evidence in favor of it is sketchy at best. Among humans, folklore and fact have long intermingled in the quest for a foolproof method to ensure the conception of a boy or girl.

Some fertility specialists say that if intercourse occurs right at the time of ovulation, then the sperm bearing Y chromosomes, known to be the fastest swimmers, are likeliest to reach the newly ripened egg first, and the child will be a boy. If, on the other hand, ejaculation precedes ovulation by a day or two, then many of the hasty Y-bearing sperm are likely to have fallen away, leaving the slower, hardier X-bearing sperm with the best chance of meeting their target and making a girl.

—NATALIE ANGIER, August 1993

The Strange Dark World of the Naked Mole Rat

IN COMPLEX BURROWS under the bricklike soil of East African deserts there lives a mammal more peculiar perhaps than any other on Earth; more peculiar, in fact, than the creatures that have emerged from the wild imaginings of Maurice Sendak or the conjurers of otherworldly visitors.

They are known as naked mole rats—though they are neither moles nor rats—and they are challenging scientists to rethink the evolution of a highly structured, cooperative way of life called eusociality, once thought to be limited to group-living insects like bees and ants. The habits of these small, fetal-looking colonial rodents, one of nine species of mole rat, are more insectlike than any vertebrate yet discovered. They are, perhaps, the best living example of what biologists call convergent evolution: the development of nearly identical characteristics in completely unrelated species.

Like termites and honeybees, the queen of a mole rat colony bears all the young, sired by only a few mysteriously chosen males. And like those of ants and bees, colonies of naked mole rats exhibit a strong division of labor. Aside from the queen and her few consorts, most of the animals are workers whose various jobs include digging to find food for the rest of the colony, taking care of the young, protecting the burrow from predators and cleaning house.

Unlike other mammals, naked mole rats do not have an internal furnace to maintain a stable body temperature. And unlike most vertebrates, individual mole rats are not engaged in a Darwinian struggle to endow offspring with superior genes. The great majority of mole rats never reproduce or even try to, sacrificing their reproductive potential to a higher good: helping the colony to survive in a very challenging environment—an act that behaviorists call reproductive altruism.

The Naked Mole Rat

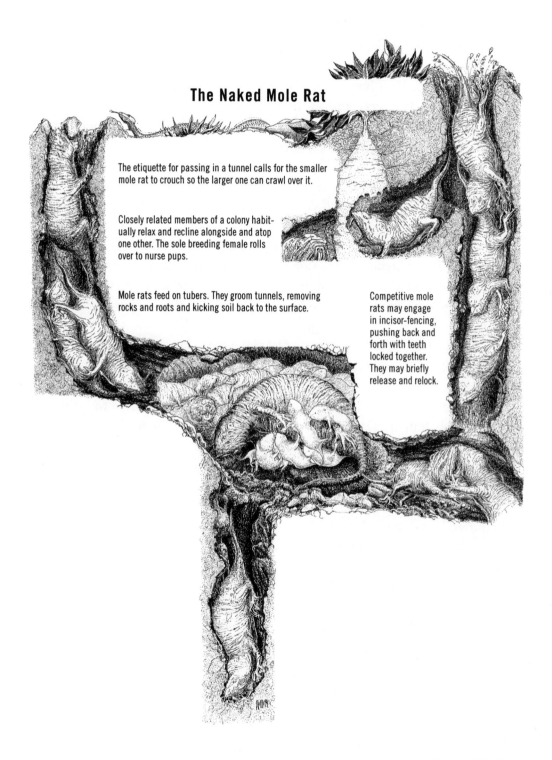

The etiquette for passing in a tunnel calls for the smaller mole rat to crouch so the larger one can crawl over it.

Closely related members of a colony habitually relax and recline alongside and atop one other. The sole breeding female rolls over to nurse pups.

Mole rats feed on tubers. They groom tunnels, removing rocks and roots and kicking soil back to the surface.

Competitive mole rats may engage in incisor-fencing, pushing back and forth with teeth locked together. They may briefly release and relock.

Dimitry Schidlovsky

In the most recent studies of this and another species of eusocial mole rat, the Damaraland mole rat, researchers at Cornell University here and the University of Cape Town in South Africa showed that these two subterranean species are forced into communal life and reproductive sacrifice by periodic droughts and patchy food supplies that make it nearly impossible for an animal to survive on its own. Solitary mole rats, on the other hand, live in ground that is easier to dig and where food is more evenly distributed.

"When the rains come, the eusocial mole rats cooperate and teams of animals dig like the fury," said Dr. Paul W. Sherman, a behavioral ecologist and Cornell professor of neurobiology and behavior. "Together, they are more likely than a solitary mole rat to find a bonanza of tubers to sustain the colony until the next rain. Alone, individuals would starve in that environment. And with a 'super mom' to produce more helpers, individuals willingly give up personal reproduction for indirect reproduction through relatives."

Dr. Hudson Kern Reeve, a sociobiologist at Cornell, recently showed through DNA fingerprinting that as a result of countless sister-brother and mother-son matings, "Mole rat colonies hundreds of animals strong are almost clones." Individuals in a mole rat colony, which typically grow about three inches long and weigh one to two ounces, are genetically more like one another than any naturally bred animals except identical twins. They are as alike, in fact, as laboratory mice that have been inbred for 60 generations.

It is this genetic similarity that allows worker mole rats to sacrifice their breeding potential, since the offspring of the queen and her mates are genetically identical to what they would be if any of the worker males had sired them, the Cornell scientists maintain.

Any harmful hidden genes that might once have been able to destroy so incestuous a colony have long since been weeded out. Dr. Stanton Braude, an evolutionary biologist at Washington University in St. Louis who has been studying about 35 mole rat colonies in the field for nearly a decade, said he had found evidence that mole rats also actively outbreed with animals from different colonies coming together to form new colonies.

Highly inbred or otherwise, mole rats are tough little creatures, surviving in captivity for as long as 16 years in plastic tunnels that serve as laboratory models of their real-world burrows.

Being so reproductively limited and living perpetually in the dark as they do, mole rats faced no pressure to evolve a certain physical comeliness that might help attract a mate or inspire popularity. Two-legged vertebrates observing naked mole rats that now dwell in about a dozen zoos here and abroad have been heard to remark, "They're so ugly they're cute!"

But in the case of the naked mole rat, "ugly" is the consequence of characteristics that have evolved to make them ideally suited for their unusual habitat. As Dr. Sherman put it in a recent interview, "When you see them alive, there's something charming, even magnificent about these animals. What we regard as ugly is really a bundle of fancy adaptations for living underground."

Dr. Reeve, a colleague and former student of Dr. Sherman, added, "The naked mole rat is a prime example of natural selection at work. That's what makes them beautiful."

If, for example, you spent your days slithering through narrow openings, you also might want loose, wrinkled skin that gives, much like the loose-fitting clothing a spelunker wears to avoid abrasions. If you could run backward as easily as forward but had teeny eyes that could see only dim shadows of light and dark, you, too, might want touch-sensitive whiskers fore and aft to help guide you along your route. You might also profit from a piglike nose loaded with chemical sensors to help you recognize your colony mates and follow scent-marked trails to sparsely distributed food.

If nature had endowed you with dainty front feet for your underground life, you, too, might want huge buck teeth to dig your way to food, not to mention lips that close behind the teeth and a fringe of hair inside your mouth to keep you from ingesting the dirt. And if you had to use your mouth for gathering groceries, you, too, might have one-quarter of your muscle mass in your jaw.

As for the lack of a hairy coat, if your abode stayed at a comfortable temperature and humidity year-round, a cloak of hair or fur would be superfluous and an ability to regulate your own body temperature would be, metabolically speaking, a costly extravagance, Dr. Reeve explained.

This mammal has the poorest known thermal regulator, and, Dr. Sherman said, "When a naked mole rat gets chilly, it simply snuggles up with its colony mates, sometimes in piles three and four animals deep. Or one animal may venture out of the burrow to bask in the warm sun and then return to huddle with the others like a living hot-water bottle."

And to compensate for the air in mole rat tunnels, where carbon dioxide accumulates at the expense of oxygen, the animals have evolved with hemoglobin that has a very high affinity for oxygen.

The fascination with naked mole rats is barely two decades old. In the mid-1970s, Dr. Jennifer U. M. Jarvis of the University of Cape Town, trying to study naked mole rats in the field, was mystified by her failure to find pregnant females in monthly samples of colonies. To simplify her studies of an animal that lives about a meter underground and almost never surfaces, she decided to capture individual animals and establish a laboratory colony. After considerable battles, particularly among the females, the colony eventually settled down and one female—and only one—began breeding.

At about the same time, Dr. Richard D. Alexander, an evolutionary biologist at the University of Michigan, was lecturing widely about the evolution of eusociality, which was then thought to be restricted to social insects like termites and honeybees. In his lecture, Dr. Alexander hypothesized that if there were a mammal that lived a eusocial life—defined by reproductive altruism, multiple generations living together, and strong division of labor—it would probably be a completely subterranean rodent that fed on large tubers and lived in a tropical region in burrows inaccessible to most predators.

At one of his lectures, another scientist informed him, "Your hypothetical eusocial mammal is a perfect description of the naked mole rat of Africa," and told him how to contact Dr. Jarvis.

The naked mole rat soon captured the interest of a handful of scientists who saw in this unusual animal an opportunity to explore how eusociality might have evolved in vastly different animals facing similar environmental pressures. Dr. Eileen Lacey, a behavioral ecologist at the University of California at Davis, said that scientists now recognize a number of vertebrates that live in a eusocial structure, among them the dwarf mongoose, African wild dog and Florida scrub jay.

"So there is not something unique about the social insects," Dr. Lacey said. "But among vertebrates, the naked mole rat is most like the social insects. No other vertebrate reaches the extremes of behavioral specialization, colony size, reproductive division of labor and morphological differences between group members exhibited by the naked mole rat."

From a laboratory mole rat colony established in 1980, Dr. Lacey said, researchers at Cornell "have long-term data on animals born in the lab and can trace their entire lives: how they've grown and behaved and how that relates to what else is happening in the colony." Enough is now known about the animal, both in the wild and in the lab, to fill a 518-page book, *The Biology of the Naked Mole Rat,* edited by Dr. Sherman, Dr. Jarvis and Dr. Alexander and published in 1991 by Princeton University Press.

But Dr. Sherman suspects that there may be still more eusocial animals. "There are lots of discoveries still to be made underground—and underwater," he said.

As a Cornell undergraduate working with Dr. Sherman, Dr. Lacey helped to determine that the division of labor in mole rat colonies was governed by body size. As with termites and honeybees, the queen is huge compared to her subjects, and she is shaped more or less like a dachshund. She is already among the larger animals when she becomes queen, and as she breeds becomes larger still.

She can bear as many as 27 offspring at a time, producing a new litter every 70 to 80 days, and pregnant or not, she is able to navigate the narrow tunnels to keep the colony in line. Should she find a lazy worker while on one of her regular patrols, Dr. Reeve discovered, she gives the ne'er-do-well a stimulating shove. And just about the time the worker may begin to slack off again, 10 to 15 minutes later, the queen is back on her rounds. Dr. Reeve also found that "the couch potatoes of the colony tend to be larger animals and less closely related genetically to the queen."

Surprisingly, it seems to be the lazy workers who grow the biggest and thus have the best chance of becoming a breeder. Other large individuals are assigned the treacherous task of defending the colony. The main threats are mole rats attempting to invade from other colonies and—most fearsome—snakes. Snakes are just about the only predators that can gain access to a mole rat tunnel and have been known to dine on as many as three or four of the rodents at one meal.

Other large nonbreeders are believed to dig for food. Smaller nonbreeders are charged with colony maintenance: transporting nesting material and food, clearing the tunnel of debris and removing obstructions.

Dr. Lacey showed that if a queen or breeding male in the colony dies or is removed, a number of other animals in the colony undergo a growth spurt, vying to replace it. And even though nonbreeding females are sterile, when the queen dies, several females suddenly grow larger, regain their sexual and reproductive prowess and battle one another to replace her. Dr. Lacey found that if a group of large nonbreeders dies, then many of the smaller animals grow faster to fill the gap. And if a colony loses a lot of small nonbreeders, the growth rate of the larger animals slows to fill that slot.

Despite some personal habits that most people would find repulsive—like ingesting their feces and rolling in the urine and excrement of colony mates, presumably to coat themselves with a common odor—naked mole rats are meticulous housekeepers. Unlike horses, they do not defecate wherever they happen to be. Rather, the colony establishes a toilet area within the tunnel system that all residents (except nursing newborns) are careful to use.

And should the smallest bit of sand or other debris drop into the tunnel, worker mole rats with maintenance responsibilities scurry to form a bucket brigade of sweepers to clean it out. The same with digging out the tunnel in search of food. One animal digs with its teeth and sweeps the dirt back with its feet, which are equipped with broomlike hairs between the toes. Then the next animal and the next and the next sweep the dirt along until it can be pushed up and out of the tunnel.

The animals also establish an underground protocol. When two naked mole rats attempt to pass each other in the tunnel, one animal—usually the smaller of the two—crouches down against the bottom of the tunnel while the larger individual crawls over it. To change directions, a mole rat uses a T-shaped tunnel junction to execute a three-point turn. First it passes the junction, then backs up into the branch, then turns and heads the other way.

A foraging animal that unearths one of the large tubers that are the mole rat's main food supply is obliged to share the cache with the colony. Studies in Cornell's colonies indicate that the animal finding food runs back to its mates, leaving a chemical trail along the route. When the researchers reversed the tunnels in the laboratory colony, the animals followed the piece of tunnel originally traversed by the food finder, suggesting they were guided by a scent rather than a directional clue. Honeybees, on the other hand, do a dance that informs others of the direction to fly in to find food.

Mole rats also use sounds to communicate, even though they have no external ears. According to Dr. Sherman, "Sound travels well in the tunnel and they have quite a complicated language, producing at least 17 different vocalizations, richest among the rodents and rivaling the vocal repertoire of some primates."

Dr. Reeve added, "We are just beginning to decipher their language. They have a distinct alarm call when threatened by a snake, for instance, and they make a hissing sound when engaged in aggressive shoving. Also, the queen emits a call in the context of mating, which may be a signal of her receptivity."

Mole rat workers make sure that nursing young stay with their mother. But once a young mole rat is old enough to leave the teat, it must get to work on colonial chores, with assignments changing according to the animal's size. Dr. Sherman suggested that young people today who come back home jobless after college might learn a thing or two from these industrious eusocial animals.

—JANE E. BRODY, April 1994

Prairie-Dog Colonies Bolster
Life in the Plains

A RISING SUN was just beginning to illuminate the harsh beauty of the Badlands when an early-rising prairie dog emerged from one of thousands of small dirt mounds decorating a swale of the Great Plains. By midmorning the ground was alive with the golden-tan, foot-long rodents, which are only now becoming recognized as an ecologically crucial species with uncommon power to affect the lives and surroundings of other creatures.

"Look at that landscape out there," said Dr. Glenn E. Plumb, the wildlife biologist for Badlands National Park, as he gazed out across the prairie-dog town. "That's a real concentration of biological activity, a lot of complex nature."

Its verdant carpet produces more new growth in a given year than is produced in similarly sized patches of the surrounding plains, attracting a crowd of plant and seed eaters, from insects to mice to birds to bison. Predators like hawks, coyotes and bug-eating birds follow.

The prairie dogs, as architects and custodians of their immediate environment, are largely responsible for this concentration of life. Their constant cropping of vegetation stimulates faster and more nutritious growth, while their burrows provide homes and hunting grounds for many organisms, large and small. The resulting patches of habitat make the Plains ecosystem more complex, diverse and biologically active than it would otherwise be.

Dozens of vertebrate species may depend on prairie-dog colonies to one degree or another, some scientists believe. At least one species, the black-footed ferret, cannot survive without them. As the prairie dog goes, conservation biologists say, so may go a raft of other creatures; save prairie dogs and you may also save a sizable and particularly valuable slice of the

Prairie-Dog World

Many species of animals and plants gravitate to prairie-dog towns. A highly social prairie dog engages in a joyous-appearing jump-yip display, upper left.

Above right;, an affectionate greeting between neighbors.

Above center, close-cropping of plants enriches vegetation available for other grazers, like pronghorn antelopes and bison. The alarm is sounded when a hungry badger appears, but burrowing owls coexist. Below, a black-tailed prairie dog flees a black-footed ferret. Rattlesnakes are another predator. Burrows must also be shared with black widow spiders and barred tiger salamanders.

Michael Rothman

Plains ecosystem. This late-blooming realization—that prairie dogs may be what ecologists call a "keystone" species—is pushing the prairie dog into the forefront of conservation concern for the first time.

Prairie dogs have long been viewed, particularly by urban dwellers, as the epitome of cute. The scene here a few days ago was typical. When the animals emerged from their dens, some touched teeth in "kisses" of recognition. They romped, rolled, wrestled and chased one another around. They assiduously tended their burrows, dirt flying from their busy paws, bundles of grass for lining underground nests drooping from their mouths.

Many stood watch, erect and alert, scanning the countryside for unwelcome guests like coyotes, hawks or people. A few executed a comically winsome maneuver called the jump-yip display: Standing upright, the animal throws its head back and its forefeet high in the air. The movement is accompanied by a sharp half whistle, half squeak that rises in pitch and then falls off again. Some experts think it is a territorial gesture, but it looks like nothing so much as an expression of joy.

Despite such charming and popular behavior, prairie dogs have for more than a century been generally regarded by humans who share their home range as noxious pests that ruin grazing land and must be exterminated. Billions have been poisoned over the decades, and the poisoning continues today with aid from both state and federal governments.

When the Sioux ruled these parts, as many as 100 million acres, or 20 percent of the land area of the Plains, may have been covered by prairie-dog colonies. Today, although two of five North American species are imperiled, hundreds of thousands of the rodents remain, which may help explain the previous lack of conservation concern. The black-tailed variety that lives hereabouts is the most widespread. Formerly endangered, it made a comeback after the most devastating forms of poison were banned in the 1970s and is no longer in jeopardy.

But conservation biologists say that the land area covered by prairie-dog colonies has been reduced by 98 percent since presettlement times. As the prairie dog's ecological role becomes clearer, this continuing loss of habitat is raising alarms about the future of less abundant species whose existence is tied to its colonies. These include the burrowing owl, which lives in prairie-dog burrows; the mountain plover, which prefers the close-cropped surface growth of prairie-dog towns; and, most notably, the black-footed ferret. The ferret, which dines almost exclusively on prairie dogs and also inhabits their burrows, is so endangered that it has had to be bred in captivity and is now being reintroduced in several places, including the Badlands.

"The trends are pretty clear, and if those trends continue, we're going to lose a lot of these species if something isn't done," said Dr. Tim W. Clark, a Yale University conservation biologist based in Jackson, Wyoming. Dr. Clark was one of several scientists who explored the subject at a meeting of the Society for Conservation Biology last month in Fort Collins, Colorado.

For all their air of fun, frolic and industry, prairie dogs—not dogs at all but a variety of ground squirrel—live in a violent and unforgiving world. They are continually beset by predators of one kind or another. The defense strategy they have evolved lies at the heart of their ecological role.

By organizing themselves in colonies and living underground, they have set up a cooperative detection-and-escape network against attack by surface and aerial predators. Any human visitor can easily see it in action, although tourists who stop at roadside pullouts on the edges of prairie-dog towns might well miss it. Prairie dogs living near the pullouts have become so habituated to humans that one can approach to within two or three feet. Biologists call these animals "Twinkie dogs," after the pastry confection to which they have sometimes presumably been treated.

But move just a little farther into the colony, and things change. Soon one prairie dog issues a warning call, a high-pitched *chirk* that it repeats incessantly, head and little black-tipped tail jerking in what looks like indignation. Others take up the call, starting a chain reaction of noise that rises in pitch, volume and speed of repetition if the intruder walks faster. Soon there is bedlam. One by one, all the animals dive into their burrows as the intruder approaches. When the danger has passed, it is high time for the jump-yip display.

The defense system is extremely efficient. "In twenty-two years, I have seen a mere twenty-two" successful aboveground predations, said Dr. John L. Hoogland, a behavioral ecologist at the University of Maryland, who has studied prairie dogs extensively at Wind Cave National Park near here. He is the author of the book *The Black-Tailed Prairie Dog: Social Life of a Burrowing Mammal* (University of Chicago Press).

But the system has its drawbacks. Predators like snakes and badgers find the burrows little or no hindrance, and lethal black widow spiders often take up residence in burrow entrances. The black-footed ferret, a nocturnal predator about half the size of a prairie dog, slips easily into burrows and exacts an especially grisly toll: It locks its teeth on the throat of its prey, strangles the victim and then eats it bones and all—except for the feet.

In some densely packed colonies, prairie dogs express a dark side of their own. In the spring, after babies are born, many mothers go on an orgy of infanticide, killing and often eating the offspring of others. They do it to reduce competition for food, and possibly also to get an extra shot of protein at nursing time, Dr. Hoogland said, adding, "It is a real slaughterhouse for four to six weeks."

Whatever the trade-offs for the prairie dogs, the big byproduct of the colony-based defense system is its impact on the Plains ecosystem, which in this region is dominated by grasses of short and medium length, as distinguished from the tallgrass prairies farther east. (Prairie dogs do not live in these eastern prairies, presumably because they cannot function well in the taller grass.)

The most obvious ecological effect arises from the burrows themselves, which attract a concentration of predators, underground-living invertebrates and other tenants like the burrowing owl. A shy bird about the size of a pigeon, it can sometimes be observed here at a distance, posted near a prairie dog on its burrow mound. (The owls do not eat prairie dogs, but batten instead on smaller prey.)

More subtle is the impact of the prairie dogs' "mowing" of the land. In foraging, as well as in creating a better view of approaching predators, the rodents clip all vegetation to within three or four inches of the surface. This changes the temperature and moisture content of the soil, encouraging some plant varieties to replace others; broad-leafed, nonwoody plants like wildflowers and legumes take over from grasses, for instance. The constant cropping also makes plants more nutritious and digestible by eliminating the decline in nutrition and roughness that come with aging. And the more rapid plant growth and recycling of energy inherent in a regimen of constant grazing simply increase the amount of vegetation over time.

"Things happen quicker in a prairie-dog colony," explained Dr. April D. Whicker, a wildlife biologist with the National Biological Service in Fort Collins, who has long studied the ecological role of prairie dogs. "Although at any one time there is less biomass, overall productivity is higher."

Naturally enough, this accelerated energy machine attracts more plant eaters, from herbivorous insects to pronghorn antelope and bison, both of which tend to hang around prairie-dog colonies in summer. The antelope are drawn by the succulent forage, while bison come both for the eating and for a place to rest and take dust baths to shed parasites.

Some studies have suggested that yearling bison that feed randomly and intermittently in a prairie-dog town gain 14 percent more weight than if they feed elsewhere. The gain is 46 percent higher, the studies suggest, if the yearlings feed regularly in colonies.

Other studies have found that the total density of animal life, apart from prairie dogs, is higher where colonies exist. Some species are especially abundant: deer mice, grasshopper mice, horned larks and some species of grasshoppers, for instance. Badgers, coyotes, swift foxes, golden eagles, bobcats and hawks are drawn to the colonies by the abundance of prey, including the prairie dogs.

While as many as 170 vertebrate species are said to appear in prairie-dog towns, it is not yet clear how many are significantly dependent on the colonies. Clarifying this picture is one purpose of a four-year study on which Dr. Whicker and two other National Biological Service scientists, Dr. Bruce Baker and Dr. Tasha Kotliar, are working.

Conservation biologists are especially concerned that although prairie dogs still exist throughout their historic range, the colonies are smaller, fewer and more fragmented. One fear is that because colonies are so widely separated and so hemmed in by farms and ranches, the overall colony system will slowly disintegrate.

Each colony has a life cycle; eventually, the vegetation plays out under intense grazing. Historically, prairie dogs simply moved to new ground, leaving the abandoned territory to recover its vegetation. Now, it is feared, they may not be able to do so as easily. And given the wide separation between colonies, it may not be possible for relatively immobile animals like the tiger salamander, which has adapted to the colony habitat, to move to a new colony.

Another concern is the threat posed by plague, which is believed to have been introduced from Asia early in this century. Once established in a colony, the plague organism, which is passed from one rodent to another by fleas, kills almost all the prairie dogs living there. A number of colonies have been ravaged in recent years, mostly in the Southwest. Since 1970, according to a laboratory of the Centers for Disease Control and Prevention in Fort Collins, 16 cases of plague in humans have been associated with prairie-dog contact. The disease responds to antibiotics.

The biggest difficulty, from the standpoint of conservationists, may be human attitudes. Though not all ranchers are unsympathetic to prairie dogs and ecological concerns, the view of the prairie dog as a pest to be exterminated is by all accounts still dominant in the West.

Cattle ranching operates on a thin economic margin, and any grazing land that is not fully productive is a liability. Ranchers commonly insist that productivity is reduced when prairie dogs compete with cattle for forage. Some conservationists argue that if the forage is sufficient for bison, it ought to be sufficient for cattle. But the jury is out on this. Dr. Whicker says that what applies in the wild, where wildlife populations are below the carrying capacity of the ecosystem, may not apply in the case of higher-density cattle herds.

National parks like this one may provide the best sanctuary for prairie-dog habitat, with other public lands second. But national forests, national grasslands and lands owned by the Bureau of Land Management are open to grazing, and much contention surrounds their use. In one area comparatively rich in prairie-dog colonies, on bureau lands in Montana, the government and local ranchers have agreed to maintain overall prairie-dog occupation at 1988 levels.

Many conservation biologists see the struggle to reach some accommodation between the interests of the Plains economy and its natural ecology as critical. Some federal land managers committed to ecosystem protection worry that congressional pressure to weaken or gut the Endangered Species Act may remove their main tool for forging such compromises. Protecting the endangered ferret, in this view, has been the key to protecting prairie-dog colonies.

"To be frank," said one land manager, "if there wasn't public land and the Endangered Species Act, we wouldn't be in the prairie-dog or ferret business. They'd be gone."

—WILLIAM K. STEVENS, July 1995

Woolly Flying Squirrel, Long Thought Extinct, Shows Up in Pakistan

MISSING SINCE EARLY in the century, the woolly flying squirrel was thought to have gone extinct, apparently having vanished from the Himalayas' frigid cliffs. But this giant among squirrels has resurfaced in northern Pakistan, setting the squirrel world abuzz and persuading many scientists that some species that have been written off may yet be hanging on by a claw out there somewhere.

"I was flabbergasted when I heard," said Dr. Charles A. Woods, a mammalogist at the University of Florida in Gainesville who is writing a book on the mammals of Pakistan. "I've worked all through there, in all sorts of high valleys in the mountains. We've really scoured the area and never seen it. This is simply marvelous."

Dr. Lawrence Heaney, head of the mammal division at the Field Museum in Chicago, said of the squirrel, which is two feet tall and sports a two-foot-long tail, "It's a spectacular animal. It's an enormous squirrel, the largest living member of the family. Just the idea of a gigantic squirrel gliding along from boulder to boulder above the point in the mountains where trees no longer grow—the rediscovery is a pretty neat thing."

The squirrel, whose cry is said to herald the death of a loved one and whose urine is claimed to be an aphrodisiac, was rediscovered not by professional biologists but by Peter Zahler, a freelance editor and writer, and Chantal Dietemann, a community college math teacher. They both live in Watertown, New York.

"It was extraordinarily exciting," said Mr. Zahler, whose freelance work supports his serious interest in natural history and whose discovery in the Sai Valley last summer has since been confirmed. "I had always thought that there were woolly flying squirrels out there." Even though the

Woolly flying squirrel, 2 feet tall with a 2-foot-long tail.

Michael Rothman

species had not been seen by scientists since 1924, Mr. Zahler said he believed it might exist because the valleys it inhabits are very isolated, with few visitors and even fewer roads.

Mr. Zahler first attempted to search for the squirrel, known as *Eupetaurus cinereus,* in the traditional fashion. Typically, a mammalogist places a heaping portion of the animal's favorite meal behind the swinging door of a trap, and waits. But difficult decisions present themselves for explorers setting out dinner for presumably extinct giant squirrels whose diet was never known in the first place.

In fact, a nearly complete lack of information on the species was the biggest hurdle. First discovered in 1888, the species has a piteously small scientific legacy consisting of a few technical papers, a single anecdotal account of a live, captive animal and a handful of specimens scattered across the globe.

Mr. Zahler tried piles of almonds, globs of honey and mounds of grain, but none lured his quarry. And after spending two months and $7,000 of his own money in 1992 and then another two months during a return trip last summer with his assistant, Ms. Dietemann, he could only conclude that if it did indeed exist, this squirrel would not be gotten by standard means.

It was only at the very end of their last trip, with almost all of their money from the World Wildlife Fund Pakistan spent, that Ms. Dietemann spotted atop a cliff a single disembodied front paw, apparently torn from the leg of a woolly flying squirrel by a predator. A few days later as the trail was growing dishearteningly cold, a pair of local men appeared and claimed they could produce a woolly flying squirrel. They said they were collectors of salagit, a supposedly aphrodisiac and medicinal substance sold in the bazaars in nearby Gilgit—and which some said was the crystallized urine of the woolly flying squirrel.

"I'd been told lots of things," said Mr. Zahler, describing his skepticism. "I'd even been told the squirrels were known to milk cows, so I didn't put too much faith in this." But the two men returned six hours later to collect their $150 finder's fee, carrying a female woolly flying squirrel in a sack.

Based on his observations, Mr. Zahler said he suspects that this creature is nocturnal, which may explain in part why the species has been difficult to observe. In addition, the collection site, where the animal was later released, was a cave high on an extremely steep and rocky slope, another explanation for the biologists' inability to find the squirrel. The collectors told Mr. Zahler that they simply came upon the squirrel asleep on a ledge in the cave and threw a bag over her.

The woolly squirrel's home was a particular surprise because most flying squirrels, including the two species in North America, make their nests in trees, never along cliffs, in caves, or in sparsely vegetated areas. Ensconced in forests, the more standard nut and leaf eaters get around by gliding from branch to branch. When they leap into the air they raise their arms, then spread their gliding membranes like a cape. The squirrels can fly as far as 100 feet.

While Mr. Zahler and Ms. Dietemann never encountered another live animal, they did collect numerous woolly flying squirrel parts, all scattered beneath the roost of a huge owl known as the eagle owl, apparently the top player in this food chain of mountain titans.

Dr. Richard W. Thorington, curator of mammals at the National Museum of Natural History at the Smithsonian Institution who has begun working with the retrieved squirrel pieces, said there is no question that Mr. Zahler has indeed rediscovered the woolly flying squirrel.

And in a burst of research on the woolly flying squirrel, Dr. Thorington, Dr. Woods and other researchers are looking into everything from where the squirrel sits in its family tree to whether the salagit found in the squirrels' caves is indeed their urine to the minute details of their wingtips. Researchers say they suspect that the species is probably extremely rare, and Mr. Zahler, who is now trying to raise money for a third trip, says he has requested endangered listing for the woolly flying squirrel from the International Union for the Conservation of Nature and Natural Resources.

Dr. George Schaller, a field biologist with the Wildlife Conservation Society, which manages the Bronx Zoo, has been trying to get reserves set up for the endangered species that live in the mountains in Pakistan along with the woolly flying squirrel, including the snow leopard. Dr. Schaller said he was delighted with the find and hoped it would bring attention to the mountains, which are quickly being deforested. "Many creatures are disappearing with nobody even knowing," he said. "So when you rediscover one, that's good news indeed."

—CAROL KAESUK YOON, March 1995

Woodchucks Are in the Lab, but Their Body Clocks Are Wild

THERE ARE NO WILD woodchucks in Australia; like Groundhog Day, which occurs every year on February 2, the plump, low-slung animals are a North American invention. But there are woodchucks at Cornell University that behave as if they lived Down Under, and others that are convinced the year is eight months long.

The mixed-up mammals are part of a continuing investigation of the species' circannual cycles by Dr. Patrick W. Concannon, an endocrinologist and reproductive biologist at Cornell's College of Veterinary Medicine. A woodchuck's level of activity, food intake, basal metabolism, body weight and amount of body fat change substantially from season to season. Dr. Concannon believes the large rodent, which has played a significant role in liver disease research, is the perfect animal model for understanding how similar rhythms can affect the physical and mental health of humans.

Recent clinical studies, Dr. Concannon said, "clearly suggest that humans also have circannual cycles, although not as profound as those in woodchucks." He added, "These cycles include changes in blood chemistry, hormone secretion, brain activity and appetite." He said that human circannual rhythms, like those of woodchucks, were most likely based on day by day changes in the length of daylight and darkness "although such entrainment may be less precise due to our exposure to artificial lighting schedules at home, work and play."

The woodchuck is found in rural and suburban areas across most of the eastern United States and southern Canada, with its close relatives, the hoary and yellow-bellied marmots, occupying western mountain regions. They are not the largest North American rodents; beavers weighing 90 pounds have been recorded. But a grizzled male woodchuck can weigh 12 pounds or more at summer's end, and the animals' size allows researchers to collect ample blood and tissue samples for studying the biological mechanisms underlying its circannual cycles, Dr. Concannon said.

A vegetarian that will eat just about anything that grows, the woodchuck, sometimes called the groundhog, is noted for its raids on gardens, its bent for burrowing in inconvenient (to people) places, and its folkloric ability to predict winter's end by emerging from hibernation on February 2 and going back to sleep for six weeks if it sees its shadow. While that date is two days short of the astronomical midpoint of winter, it is too early for woodchucks to wake up in northern parts of their range like New York State, Dr. Concannon said. "The majority of males come out in mid- to late February and females appear from late February to the middle of March," he said.

Dr. Concannon described the period from mid-March to May as a time of ravenous appetite and hyperactivity as woodchucks rush to mate and raise a litter by June so the young will have time to prepare for hibernation. "Woodchucks born after the end of April probably won't survive the winter," he said. "That's why the breeding season is early and short."

He said that while the food intake of laboratory woodchucks increased twentyfold in the spring, "just as in the wild, there is little weight gain. The animals are hard-pressed to keep up with their energy expenditure and a hundred percent increase in their metabolic rate."

In early June, however, woodchucks' metabolism slows, and while their food intake also decreases, the animals increase their weight as much as 100 percent during the summer. Most of their energy goes to producing the fat deposits on which they will live during hibernation as well as in late winter when fresh grass is unavailable, Dr. Concannon explained. Woodchucks will stop eating altogether by September, he said, and they will be asleep in their burrows by mid-October. Unlike bears, which he called "pseudohibernators" whose body temperature in winter remains near normal levels, woodchucks allow their temperature to fall to 40 degrees. "They would freeze if their burrows were not below the frost line," the scientist said.

Dr. Concannon's studies, published in several scientific journals, are conducted at Cornell's woodchuck laboratory on the outskirts of Ithaca, New York. There, in comfortable digs with stovepipes for burrows, the world's only disease-free colony of several hundred of the rodents is maintained for medical research. Wild woodchucks are often infected with a virus very similar to human hepatitis B, and among other discoveries, Cornell scientists have shown that immunization against the virus can prevent the liver cancer that normally develops in infected animals.

Since other scientists had determined that the seasonal rhythms of woodchucks did not change when they were constantly exposed either to 16-hour or 8-hour days, Dr. Concannon decided to experiment with computer-controlled lighting that mimicked nature, with gradually longer and then shorter periods of daylight. He synchronized one group of laboratory woodchucks to a circannual cycle as if they were living in Australia, breeding during the Southern Hemisphere spring, and another group to an eight-month year.

Dr. Concannon said his findings showed that woodchucks had powerful hormone-driven cycles that were synchronized by changes in the photoperiod, even when the animals were hibernating in their burrows and there are no light cues. "The cycles are so strong that in the laboratory, where we maintain the temperature at seventy degrees year-round and the animals have ample food and water, some woodchucks not only stop eating but also try to hibernate," he said. "They lower their body temperatures to about seventy-five degrees, which would kill humans and most other mammals."

Dr. Concannon said the testes of male woodchucks, which shrink during the shorter days of autumn, spontaneously redevelop late in the hibernation period. He theorizes that the effects of the sleep-inducing hormone melatonin, which the animal's system secretes continuously in the darkness of the winter burrow, wear off after three to four months. "In effect, the woodchuck gets an internal message that says, 'Whoops, it's just about time to wake up and breed,'" Dr. Concannon said.

—LES LINE, January 1997

3

THE CHARACTER OF CATS

The mammalian order of the Carnivora has some 10 or so families, of which the most predatory and fearsome are the Felidae, or cats.

With bodies designed for hunting, their typical strategy is to stalk their prey, then surprise it with a swift leap or chase. Few other animals so well combine beauty with terror, cruelty with elegance.

When early prehuman primates left the forests for the African savannahs they may have had much to fear from these competitors. Several early hominid skulls bear distinctive double puncture marks that match the span of a leopard's jaws. Strangely enough, though, people lack the instinctive, deep-rooted fear of large cats that they harbor for snakes.

Cats are the essential predator. Their bodies are specially adapted for the pursuit and devouring of prey, mostly other mammals. They have keen hearing, sharp sight and sensitive whiskers to help them move noiselessly through undergrowth. They possess powerful hind legs for springing and a recessed collar bone that doesn't snap when the forelegs collide with the prey.

Cats tend to swallow meat unchewed, and because of this habit have fewer and smaller cheek teeth than other mammals. That is why their faces are shorter, flatter and more humanlike.

The living families of carnivores are divided into two major groups, the catlike and doglike superfamilies. In the catlike superfamily belong the Hyaenidae (hyenas) and the Viverridae (civets) as well as the Felidae (from house cats to lions).

Survival of the Big Cats
Brings Conflict with Man

THE TIGER AND THE CHEETAH appear to be on their last legs. The grizzly bear and the wolf are struggling desperately to hang on in the Rocky Mountains. But biologists say the resourceful and mysterious cougar, alone among the world's big endangered carnivores, has sprung back spectacularly from the brink of extinction and now reigns as the king of wild predators in North America.

As cougars repopulate much of their former range, however, they are also encountering humans more frequently. Sometimes there are tragic consequences, as when, in April, one of the big cats ambushed and killed a mother of two children in Northern California. The incident raised with special intensity the issue of whether and how expanding populations of cougars and people can coexist.

Cougar, puma, mountain lion, panther, painter, catamount: By any name, the biggest of the purring cats (as opposed to roaring cats like tigers, leopards and African lions) epitomizes grace, stealth and steel-muscled power. The tawny hunter with the black mustache and long tail has so captured people's imagination across the centuries that many Indian tribes revered it as a god and today it is the icon of team after team.

The fascination did not stop European settlers and many of their descendants from trying to exterminate the breed. Once the cougar inhabited the largest territory of any land mammal in the Western Hemisphere, stretching from the Yukon to the Strait of Magellan and from the Atlantic to the Pacific. But bounty hunting drove the species nearly to the vanishing point in North America by the mid–20th century. Now cougar hunting is

Temporalis
muscle

Masseter
muscle

Carnassials

Canines

A Hunting Machine Makes a Bold Comeback
The cougar, or mountain lion, also called the panther or puma, the largest of the purring cats, is stronger pound for pound than other big cats and is very well adapted for stalking, ambushing and killing prey in rough terrain.

High-Power Jaws
The short round skull, with huge temporalis and masseter muscles, can apply tremendous force through the jaws. The large canines are used to sever the prey's spinal cord with a bite to the back of the neck. The rear teeth, called the carnassials, shear large chunks of meat into strips, which the cat swallows whole.

Reclaiming Lost Ranges
At this scale the map for the current range does not show populations in Florida and possible small populations in Manitoba and New Brunswick, Canada. The status of the cougar in Mexico and Central and South America is not known. The last cougar in Vermont was killed in 1881; the last one in Pennsylvania was killed in 1891.

Michael Rothman

restricted in all states from the Rockies to the Pacific. Limits on hunting deer and elk, coupled with milder winters, have brought about a population explosion among the cougar's main prey. As a result of all these factors, experts say, mountain lions have become common once again throughout much of the West.

The resurgence "is a raging success story," says Dr. Maurice Hornocker, a mammalian ecologist who has been studying cougars since the 1960s. He is director of the Hornocker Wildlife Research Institute, a private organization affiliated with the University of Idaho in Moscow. "The mountain lion, without any help from special committees, commissions, congressional action or anything, has come back from very, very low numbers," he said. And while only educated guesses about cougar populations are possible, he said, the cats may be more abundant now than when the West was beyond the frontier. Estimates place cougar numbers at 9,000 to 12,000 in just California, Colorado and Idaho combined.

The problem is that as the cougars have been rebounding, their natural territory has been increasingly occupied or routinely visited by an even more expansive human population. More cougars plus more people equals more encounters between the two, including more attacks on people, though such attacks remain rare. Dr. Paul Beier, a wildlife ecologist at Northern Arizona University in Flagstaff, has documented 57 cougar attacks since 1890, 43 of them since 1970. Of the total, 11 attacks were fatal and 7 of those occurred after 1970—2 in the last four years, one each in Colorado and California.

By contrast, automobile accidents involving deer kill scores of people every year, and more Americans die of bee stings or lightning bolts than cougar attacks. But western wildlife officials say the odds in favor of a cougar encounter are much higher in many western localities, and the horror of a fatal attack, however rare, sends out shock waves. The mountain lion is in one way the most fearsome of all the big cats; it can kill bigger prey, relative to its own size, than any other. Those steel muscles allow it to drag down a bull elk seven or eight times as large. An adult human, who weighs about as much as a cougar, is no match.

So it was on April 23 when Barbara Schoener, a 40-year-old long-distance runner, vocational rehabilitation counselor, wife, and mother of two preteen-age children, was running for relaxation alone in a recreation area 45 miles northeast of Sacramento, California. As reconstructed by state wildlife officials, Mrs. Schoener's course took her along an isolated trail on a hillside. A mountain lion sat above the trail, presumably on the alert for deer.

"The lion was doing what a lion would do," said Bill Clark, senior wildlife biologist in charge of the California state wildlife investigations laboratory at Rancho Cordova. When Mrs. Schoener ran by, he said, "it was like rolling a ball by a kitten." The cat apparently attacked the runner from behind—cougars, relying on stealth and surprise, usually ambush their prey from the back or side—with such force that Mrs. Schoener was knocked off the trail and down the hill.

Cuts on Mrs. Schoener's hands suggested that she might have struggled briefly. Death probably came quickly. Cougars usually kill by inserting their teeth between the victim's neck vertebrae and severing the spinal cord. Mrs. Schoener was bitten on the head and neck, and an autopsy revealed that either bite could have been fatal. The cat ate part of the body and buried the rest for safekeeping under a pile of sticks and leaves, just as it normally would bury a deer. The remains of several lion-killed deer were found similarly hidden in the vicinity.

Mrs. Schoener's death brought angry calls from hunters, ranchers, farmers and some legislators to repeal a 1990 California law that bans cougar killing. The only exceptions to the law are in cases of lions that actually threaten or attack humans or livestock. (California is the only western state that prohibits hunting the cats.)

Conservationists, unwilling to jeopardize such a spectacular victory as the recovery of a major mammalian species, were just as vocal. They have long argued that protecting cougars is important because of their ecological role at the top of the food pyramid and because countless other species will automatically be protected, as well: Cougars need home ranges of up to hundreds of square miles, and protecting those ranges necessarily protects many other creatures, big and small, together with the ecosystems in which they exist.

The effort to repeal the ban on hunting cougars has died in the California legislature, at least for now. Public opinion in favor of the cougar may have been partly influenced by the discovery, after wildlife officials tracked down and shot the animal believed responsible for Mrs. Schoener's death, that the lion had left a cub.

Despite the occasional attacks, many people seem favorably disposed toward cougars. The public's "tolerant attitude" is even more striking than the lion's strong comeback, Dr. John Seidensticker, the curator of mammals at the National Zoological Park in Washington, D.C., wrote in 1992 in *Smithsonian* magazine. Dr. Seidensticker spent several years studying cougars with Dr. Hornocker. In terms of public interest and excitement, "the mountain lion is up there with whales and baby seals," said Mark Palmer, the executive director of the Mountain Lion Foundation, a research and advocacy organization based in Sacramento, California.

Beauty, intelligence, power, independence, symbol of a vanishing wilderness—all these attributes are invoked by puma partisans as reasons many people respond to mountain lions so favorably and with such awe. But mystery may also account for much of whatever fascination people feel.

Cougars are solitary and hard to find. "Even when you're studying them you rarely see one," says Dr. Beier, who has studied mountain lions in California. A cougar's radio collar can indicate that it is no more than 20 or 30 feet away, he said, but the animal can still be invisible.

This secretiveness, central to the cougar's predatory lifestyle, also enabled it to withstand, if narrowly, a centuries-long, continentwide effort to exterminate the species (the effort succeeded in most of the East). Cougars' elusiveness also made them all but impossible to study until Dr. Hornocker first used tranquilizers to subdue them in the 1960s and radio collars to track their travels.

Much has been learned about them since then. *Felis concolor,* as the species is known scientifically, consists of some 26 subspecies, or geographic races, scattered across North and South America. They have demonstrated the species' superior adaptability by mastering habitats as diverse as high mountains, Pacific rain forests, southwestern deserts and the Florida Everglades. The two eastern subspecies, the Eastern and Florida panthers, are classified as endangered. Cougar populations in many suburban areas of the West are facing critical habitat loss, and one western subspecies, the Yuma puma, is a candidate for the federal endangered list.

It is on large stretches of undeveloped public and private land in the West that the lion has flourished of late. The range of a lone female cougar often overlaps the ranges of males and other females. The much larger male ranges seldom overlap, and the residents patrol them jealously. Generally, a lion occupies its range for life if it contains sufficient prey.

When not hunting, males spend most of their time searching for mates. Mating takes place in the female's home range and consists of as many as 50 to 70 caterwauling acts of copulation a day for about a week. Scientists believe this intense mating's purpose is to stimulate ovulation.

A female bears one to six spotted kittens. She teaches them hunting and survival skills through example, and they seek their own solitary existence within a year or two. These "transients" wander, homeless, until a resident cougar dies, vacating its range, or is driven off by direct challenge to become a transient itself.

Biologists believe some transients head for suburban areas, where prey pickings, including family pets, are easier. Juvenile transients, inexperienced as they are at catching wild prey, are also believed to be responsible for most attacks on humans. (The lactating adult thought to have killed Mrs. Schoener was apparently an exception.)

The biggest remaining gap in knowledge about cougars may be in precisely the area of most immediate concern: contact between mountain lions and humans. "I think we know very little about what's actually going on in terms of these interactions with people," said Dr. Seidensticker.

Prohunting advocates argue that restrictions on hunting have caused cougars to lose whatever fear they have of humans and become more aggressive in attacking both people and livestock. The answer, these advocates argue, is liberalized hunting policies.

Dr. Hornocker says studies of radio-collared cougars provide strong evidence that some have indeed become habituated to people. "We know they can come very close to human activity without being detected," he said, "and in some regions we find that some lions now appear to have no fear of humans."

Mr. Clark says, however, that California studies of collared lions show that while they may approach people with no fear, they do not necessarily bother them. "We've observed lions observing people for long periods of time, months and months, without being aggressive," he said.

But occasionally they do seem menacing, as was the case last Thursday in Lockwood, California, when a cougar wandered a residential neighborhood for two hours and then perched in a tree near a schoolyard, watching children play. State wildlife officials decided this behavior was too aggressive and the cat was shot. Five to 10 cougars are killed in California for similar reasons each year.

Dr. Beier rejects the idea that hunting deters cougar attacks because Vancouver Island, where the cats are heavily hunted, experiences the most attacks by far. Dr. Hornocker theorizes that some subpopulations of cougars may innately behave differently toward humans. On Vancouver Island, where the cougar population is genetically isolated, "we literally have a different breed of cat," he said. Before laying down rules for cougar management in any given locality, he said, it is necessary to understand the broad spectrum of cougar behavior toward humans from one region to another.

Killing cougars habituated to humans would indeed solve the problem of attacks on people, Dr. Hornocker says—if the right cats could be identified. But hunting, he said, is indiscriminate: "You could remove an animal that's no threat to humans, making way for an animal that could become a threat."

While waiting for more precise advice from scientists, many wildlife managers in the West are concentrating on educating the public about how to behave in cougar country. For instance, biologists say, running in the presence of a cougar is about the worst thing one can do; it stimulates the lion's instinct to chase.

"Many peoples of the Earth have learned to live for centuries with big, dangerous animals," says Dr. Hornocker. "People in Montana, in Canada and Alaska who have lived for a long time with grizzly bears have had very few problems because they know how to behave, what to do and what not to do in bear country. We need to do the same thing with mountain lions."

THESE are among the dos and don'ts of behavior in cougar country, as recommended by wildlife officials in California and Colorado:

- Do not feed deer, raccoons or other wildlife, or leave out garbage; these animals are potential prey and can attract cougars.
- Do not feed pets outside. The food attracts wildlife.

- Avoid planting vegetation on which deer might feed. State wildlife officials can offer advice on how to "deer proof" property.
- Remove dense and low-lying vegetation where cougars can hide.
- Install outdoor lighting.
- Keep pets in the house or in a kennel. Some cougars may prey on them.
- Where practical, keep livestock in enclosed sheds and barns at night.
- Keep a close watch on children whenever they play outdoors. Cougars and other big cats are excited by them.
- Keep children close by when hiking.
- Do not hike alone. Go in groups.

If a cougar is encountered:

- Do not approach the animal. Most will try to avoid a confrontation.
- Do not run. Running may stimulate the cat's instinct to chase.
- Do not crouch or bend over. An upright human is not the right shape for a cat's prey, while one bending over looks much like a four-legged prey animal.
- Do all you can to appear larger. Open your jacket. Make eye contact, wave your arms slowly and speak firmly in a loud voice. Throw stones, branches or anything you can reach without crouching or turning your back. The idea is to convince the cougar that you are not prey.
- If attacked, fight back with whatever is handy. Even caps, jackets or bare hands can be effective. Since a cougar usually tries to bite the head or neck, try to remain standing and face the animal.

—WILLIAM K. STEVENS, August 1994

Please Say It Isn't So, Simba: The Noble Lion Can Be a Coward

THE SPECTACLE IS becoming all too familiar. One by one heroes are being knocked from their pedestals, their veils of nobility, bravery and om-
nipotence stripped way to reveal a selfish, cowering heart. Now, to the
legion of fallen sports stars, artists, politicians and religious leaders must
sadly be added the cream of cats, the universal symbol of courage and royalty, a beast powerful enough to match Mickey as one of Disney's biggest
money machines: the lion.

Where once they considered the lion to be a model of cooperative behavior and unflinching boldness, scientists have discovered that some lions can in fact be malingering cowards, Bert Lahr with no Wizard of Oz in
sight. The new findings challenge conventional theories about how cooperation among social animals evolved, and suggest that life out there in the
wilderness is even nastier than anybody thought.

Reporting in the journal *Science,* Dr. Robert Heinsohn of the Australian National University in Canberra and Dr. Craig Packer of the University of Minnesota in Minneapolis demonstrate that there are basically two
types of lions in any given pride: one that will go forth to challenge intruding lions and keep the territory safe for its fellows, and another that will try
to shirk its guardian duties, hanging back as long as possible and leaving
the intrepid ones to risk their hides in pride defense.

"It turns out that lion-hearted lions and cowardly lions coexist," said
Dr. Packer. "The majority of lions in a pride seem to be good citizens, but
they support a small proportion of freeloaders and tax evaders."

The results contradict standard theories suggesting that animals living
in a group adhere to a so-called tit-for-tat strategy, cooperating when its co-

horts cooperate, and punishing those who try to cheat and either get more than their fair share or fail to perform group tasks.

In this case, the lions who lead the defense recognize a laggard as a laggard, but they do nothing to punish the offender. They do not evict the coward from the pride, nor do they try to turn the tables the next time around and hold back while the laggard goes forward to meet a threat. Leaders are consistently leaders and chickens are always chickens.

And while a position of leadership confers benefits on some species of animals, like extra food or the best mates, in the case of lions a real egalitarianism reigns. Lions share the big game they catch, they nurse communally and they have no apparent hierarchy. The leaders are not dominant over the laggards; they are simply more willing to fight for their territory, a critical behavior if a pride is to keep its turf and thus its communal life.

Researchers are flummoxed about how such a cooperative system could have evolved. Normally, natural selection operates by the laws of pure selfishness, but in this case, the leading lions seem to be risking their lives for the laggards with no apparent payoff. Scientists do not yet know whether the cowardly lions make up for their desultoriness in some other way—say, by being better hunters or better milk producers—or whether the laggards in fact are sponging off the majesty of their peers.

Moreover, the dynamics of laggardhood are surprisingly complex. Some of the laggards will shirk their responsibilities no matter what. A second type only stays behind when the threat is minor, that is, when there are but one or two intruders; these cats will rouse themselves to action when the number of interlopers grows to three or four.

Other laggards are all gung-ho when the threat is modest, but shrink when the danger to the pride grows deadly.

"This is what I found so exciting about the study," said Dr. Packer. "The lions are sensitive to the odds. There are conditional heroes and conditional cowards, just like the real world."

Dr. Lee Alan Dugatkin of the University of Louisville in Kentucky, who studies cooperation among animals, said, "I thought it was interesting that given all the theoretical models of cooperation out there, in fact the lions seem to be doing something that none of the theories account for. It's nice that instead of theoretical work driving empirical work, this will make the theoreticians sit down and ask, How does this system evolve?" In other words, nothing ruins a good idea like the real world.

Dr. David W. Stephens, a behavioral ecologist at the University of Nebraska in Lincoln, said, "This is one of a growing body of papers that suggest we need to take a more pluralistic view of how cooperation in animal societies is organized."

The scientists performed their experiments on eight prides in the Serengeti National Park and Ngorongoro Crater in Tanzania. With the help of an audio system, they focused on the female response to danger. Among lions, males keep out intruding males while females chase away alien lionesses. Lions are deeply territorial and declare their home as their own by roaring. "The really evocative sound of the African night is the territorial roar, which says, 'This land is my land,' " said Dr. Packer. "Lions don't roar unless they're very comfortable where they are, and they will fight tooth and claw to keep it for themselves."

The scientists recorded such territorial roars from different prides and then played them back to other prides, using tapes of one lion or of two, three or more lions roaring in tandem. In some cases, the researchers placed a stuffed lion near the roaring speaker.

"Judging by everything we can see, the playbacks are incredibly realistic to the lions," said Dr. Packer. "They respond to the speakers, and they'll attack the stuffed lion as though it were a real lion—though afterward they look embarrassed about it."

Through the playbacks, the scientists determined that some lions lagged considerably behind others in approaching the counterfeit enemy, sometimes reaching the target several minutes after the leaders had gotten there. Several minutes is a long time in a lion war, long enough for the leader to have been attacked and possibly killed.

The biologists then began refining the experiments by pairing leaders with other leaders, or leaders with laggards. In each case, the pair was temporarily separated from the pride. Sure enough, the leaders consistently met the challenge together, while the laggards consistently lingered in the rear. Significantly, the leaders realized when they were paired with slowpokes, for they would hesitate in their forward march, look back over their shoulders as though to say, "Come on!" and yet still continue in the vanguard. "The leader would arrive at the speaker several minutes before her companion," said Dr. Packer. "She did not refuse to carry out the response against the stranger, even when paired off with one of the jerks."

By varying the number of roars in the playbacks, the scientists observed that the laggards changed their tactics in different ways, depending on whether the menace was large or small.

The results cannot be fitted into current game theories, mathematical models that try to predict group dynamics in nature. Most of the models assume some form of reciprocity, that the animals respond to one another in a familiar eye-for-an-eye fashion. "These lions are obviously capable of some individual recognition, and they go through significant cognitive processing," said Dr. Dugatkin. "But in fact they don't use this for score-keeping or reciprocity."

Dr. Luc-Alain Giraldeau, a behavioral ecologist at Concordia University in Montreal, believes that the laggards are essentially behavioral blood-suckers. He said that in general the lions must cooperate to defend their territory, or all of them will die; but as long as the majority of cats are lion-hearted, the system is susceptible to small amounts of parasitism, in the form of a few who do less than their share.

In Dr. Packer's view, the fact that the lions do not retaliate against cheaters is grim news for those who nursed hopes of discovering at least one shining example of workable animal communalism; essentially, he said, the game of life seems to be, Steal what you can get, give as little as you can and cheat at every opportunity—even when you are among friends.

"The absence of reciprocity tells us the world is really vicious out there," he said. "Lions are meant to be a paragon of cooperation, but if they can't manage reciprocity, what hope is there for cooperation in the jungle?"

—NATALIE ANGIER, September 1995

Lions Find Peace Without Rankings

For most group-living species, the definition of "social" means knowing your place and showing it at every opportunity. In rare cases, however, the assembled animals do not bother forming a dominance hierarchy, preferring instead a form of relaxed anarchy. The most outstanding example of a species that lives in a stable group without bothering to assign rank to its citizens is the lion.

Within a pride, the resident males clearly dominate the females, and the adult females in turn take precedence over the cubs when it comes to divvying up the prey. But the great cats do not establish class systems along gender lines, as social animals usually do. If there are multiple males in the pride, they all behave like lazy, surly, demanding princes; while the females, who may number 20 per group, live in a democratic sisterhood, with no single lioness displaying pretensions of royalty.

"It's still completely astonishing to me that we can find virtually no evidence of dominance ranking in any pride," said Dr. Craig Packer.

In fact, when Dr. Packer and his wife, Dr. Anne Pusey, also at the University of Minnesota, first started studying lions in the Serengeti region of Africa in the 1970s, they were told by other scientists that the carnivores did not form any obvious class system, but they refused to believe it. After years of watching primates flaunt their social status, "we thought there has to be rank" among lions, said Dr. Packer. They soon realized that they were wrong, and they have since sought to understand how the lions keep the peace among themselves without a top-down structure, particularly during high-stress situations like breeding and feeding.

It turns out that lions adhere to what the renowned evolutionary biologist Dr. John Maynard Smith of the University of Sussex in England calls the "bourgeois strategy," the principle that whoever gets to a piece of property first—be it meat or a mate—is the de facto owner of that property. For example, if a group of lionesses pulls down a wildebeest, the lioness closest to where the best pickings are, the viscera near the anus of the carcass, essentially "owns" that neighborhood, and gets to eat her fill there. When she is finished and waddles off, the lioness nearest her old spot will move in to finish whatever is left. Dr. Packer and Dr. Pusey have noticed that if female A was at the choicest part of the prey after one kill, female B or C will be at the best table the next time. Similar rules of proximity dictating possession apply when it comes to mating.

Why should lions live by egalitarian rules when most other social creatures do not? Dr. Packer suggests that lions are simply too deadly and well-armed to manage otherwise. When two baboons or chimpanzees get into a fight, the victor usually emerges with only a few scratches and maybe a bit of ear gnawed away. But if two lions start quarreling, even a younger, smaller animal can inflict severe if not fatal wounds on her older, bigger antagonist. "A lion's got claws that can disembowel a zebra, as well as those god-awful teeth, so an inferior could cripple a better opponent," said Dr. Packer. "It's like a cold war, with mutually assured destruction. It's better not to mess around with your opponent in the first place."

—NATALIE ANGIER, April 1995

Cheetahs Appear Vigorous Despite Inbreeding

THE CHEETAH MAY BE a gorgeous Maserati among mammals, able to sprint at speeds approaching 70 miles an hour, yet it has not been able to run away from its many miseries.

Once the cat ranged throughout the African continent, the Near East and into southern India; now it is extinct almost everywhere but in scattered patches of the sub-Sahara. Farmers and ranchers in Namibia shoot them as vermin. On reserves, where cheetahs are often forced into unnatural proximity with other predators, they are at the bottom of the meat eaters' grim hierarchy; lions will go out of their way to destroy cheetah cubs, while hyenas, leopards and even vultures can easily chase away a cheetah from its hard-caught prey.

And to make the magnificent cat's story more poignant still, many scientists have concluded that the species is severely inbred, the result of a disastrous population crash thousands of years ago from which the poor beast has hardly had a chance to recover.

Studies of cheetah chromosomes have shown a surprising lack of genetic diversity from one individual to the next, and as a result the cheetah has been widely portrayed as sitting under an evolutionary guillotine, the population so monochromatic that, in theory, a powerful epidemic could destroy many if not all of the 15,000 or so cheetahs that survive in the wilderness.

Some zoos have complained that their cheetahs are infertile, and they have attributed the problem to the cheetah's bleak genetic makeup, calling into question the long-term prognosis even for cats living in the pampered confines of a park.

Now scientists at the Center for the Reproduction of Endangered Species at the Zoological Society of San Diego argue that this widely held notion of the inbred cheetah may be way off the mark, if not outright wrong. They insist that, far from displaying the negative effects of inbreeding seen in other animals known to be genetically homogeneous, like some strains of laboratory mice or pedigreed dogs, cheetahs are in many ways robust, more like ordinary house cats than the feeble product of what amounts to generations of incestuous couplings.

But the significance of the debate extends far beyond the spotted greyhound of a cat. Scientists are now seeking to calculate the odds that any number of endangered or threatened species are likely to survive into the 21st century, and among the many questions they are asking is how much genetic diversity a creature requires if it is to rebound from the brim of extinction.

Inbreeding is thought to be harmful to a species for two reasons: First, because it allows hazardous recessive traits that are normally in the genetic background to come to the fore, resulting in birth defects, still-births and in some cases infertility; and second, because it leads to a genetically uniform population without the diversity to resist epidemics and environmental changes. But the San Diego scientists said their cheetahs almost never bore defective cubs, were perfectly fertile and had great variation in their immune systems.

"Cheetahs have been likened to inbred lab mice," said Dr. Donald G. Lindburg of the San Diego Zoo, who heads its cheetah breeding and research program. "This is the dogma that is so entrenched in the scientific literature right now. But when I think of inbred mice, I think of reduced litter size, retarded growth, retarded vitality and congenital defects. None of these problems apply to the cheetah as we know it."

Dr. Lindburg and a colleague, Dr. Michael B. Worley, a virologist and immunologist, argue that while the cheetah may look genetically tenuous when its DNA is appraised, by such real-life measurements as fecundity, litter size, cub health and immune response, the cat is perfectly fit for the next millennium.

The work calls into question the validity of taking a strictly molecular approach to the sometimes murky science of species preservation, and it strongly suggests that scientists do not yet know enough about how certain genetic patterns detected in laboratory tests translate into the genuine strengths and weaknesses of a wild animal.

Those zoos that have trouble propagating cheetahs in captivity, said Dr. Lindburg, should not blame the animal's DNA, but rather their own ineptitude at animal husbandry and matchmaking. At the San Diego Zoo, the cheetahs breed so readily that the keepers sometimes call a moratorium on reproduction, just to keep down the resident cub population.

Others sharply dispute the conclusions of the San Diego scientists, maintaining that cheetahs are indeed abnormally inbred and that their genetic monotony does compromise their long-term prospects. They said cheetahs clearly suffered from chronic health problems that might be linked to a defective immune system and overall genetic frailty.

The San Diego scientists "are trying to attack what has become a commonly accepted series of experiments," said Dr. Stephen J. O'Brien of the National Cancer Institute in Frederick, Maryland. "I've written ten papers on the genetic structure of cheetahs, and in every case we've been able to support the thesis that the genetic structure is remarkably depleted compared to other big cats."

Dr. O'Brien, who is considered the world's authority on molecular studies of exotic cats and other endangered animals, was the principal author of the first and most spectacular report alerting the world to the cheetah's genetic plight, a paper that appeared in the journal *Science* almost 10 years ago.

He performed experiments demonstrating, for example, that when skin from one cheetah is grafted onto another, it takes an extraordinarily long time for the immune system of the transplant recipient to reject the added flesh—strong evidence that cheetahs are practically clones of one another.

Through extensive DNA analysis, Dr. O'Brien has concluded that the cheetahs suffered a population crash 10,000 years ago, at the end of the last ice age. He postulates that humans, advancing rapidly in the wake of the retreating glaciers, in short order wiped out the creatures everywhere but in pockets of Africa.

In that mass extermination, he said, cheetahs lost more than 90 percent of their genetic variation, and they have since managed, through gradual mutations in DNA, to recover only a fraction of the diversity.

If Dr. Lindburg and Dr. Worley have such great evidence to contradict the premise of cheetah inbreeding, Dr. O'Brien said, they have yet to persuade him.

"I have a less than enthusiastic reaction" to their latest arguments, he said. "I like Don and Mike, but if their work was really any good, they would be dying to get my approval. They don't seem to be doing it."

Dr. Linda Munson of the College of Veterinary Medicine at the University of Tennessee in Knoxville, who studies cheetah diseases, suggested another explanation for the cheetah's depleted DNA. She proposed that the cat lacks genetic diversity not because it once suffered through a catastrophic population bottleneck, but because it is the most specialized cat of all, with a body designed from snout to spine for the sole purpose of running at supermammalian speeds.

By this argument, the evolutionary process that focused on enhancing the cat's capacity to sprint ended up throwing out a lot of other genes along the way. In other words, the business of being a cheetah could require genetic homogeneity, and a modest life span could be part of the package deal.

And the cheetah is a spectacular example of streamlined design. It is relatively petite and light-boned, weighing only about 70 pounds, has an aerodynamically small head, unusually long legs, a flexible spinal column and a sliding shoulder to lengthen the stride.

Its canine teeth are very small to leave plenty of room for its nasal passages, which are extremely wide so the animal can take in a lot of oxygen. The cheetah hunts not by stalking prey, but by bolting at its quarry in an explosion of energy so exhausting the cat has to wait 15 to 20 minutes, panting, before it can eat.

Because the cheetah is slighter than most other African carnivores and lacks large canines with which to defend itself, it cannot ward off competing meat eaters that want its dinner, and when confronted it will usually give up and skulk away. Indeed, the cat is so unaggressive by nature that when a visitor went into a large pen at the San Diego Zoo to see a mother and her five young cubs, the cats allowed her to approach to almost within stroking distance, the mother looking on with a mixture of boredom and irritation, the cubs cutely raising up their fur and hissing ever so slightly.

Some ecologists see the cheetah's long-term future resting not on genetic research but on old-fashioned remedies like preserving its remaining habitat and collaborating with Africans. In Namibia, for example, where the cheetah does not have to compete with many other carnivores, as it does in other parts of Africa, the feline is faring reasonably well, and its biggest problem is that some ranchers shoot it in misguided defense of their livestock. Biologists working there are seeking to persuade the cattle owners that cheetahs kill very few livestock animals.

"Namibia is one of the last countries with a big free-ranging population of cheetahs," Dr. Munson said. "If the ranchers can be convinced to take pride in the cheetah as a unique Namibian species and to work for their survival, this may be the cheetah's best hope."

—NATALIE ANGIER, November 1992

Medicinal Potions May Doom Tiger to Extinction

WHEN THIS REACH of dry forest and lotus-covered lakes in central India was a private hunting reserve of the Maharajah of Jaipur, the beaters' cries of "Bagh! Bagh!"—Tiger! Tiger!—sounded the death knell for the animal considered by many as the greatest of the cats. The tigers that fell here to the hunters' guns were among tens of thousands "bagged" during the period of British rule in India alone.

Now, in the forest stillness, the shrill calls of the langur monkeys and the sambar deer are what signal a tiger's approach. But the calls are increasingly rare, for Ranthambhore's tigers, like all India's tigers in the wild, are threatened more than they ever were in the era of the maharajahs and the British sahibs who made sport of shooting tigers from hunting towers and the safety of elephants' backs.

The Bengal tigers that roam India's forests and grasslands, and the five other surviving tiger species elsewhere in Asia, are in danger of becoming extinct. At the turn of the century, after at least a millennium of tiger hunting, perhaps 100,000 tigers remained in the wild, ranging across a vast triangle from the Caspian Sea in the west to Sumatra in the east, and to Siberia in the north. Now, wildlife experts believe there may be fewer than 5,000 tigers left, two-thirds of them living with growing precariousness in India.

After two decades of official assurances that tiger populations in India's reserves were on a healthy rebound, a series of poaching scandals, starting here in Ranthambhore two years ago, has prompted the government to declare a "tiger crisis" and promise urgent action to save the animal that serves as a national symbol. For India, where the tiger has been alternately worshiped and feared for millennia, and where it serves as a major tourist attraction, the realization that the tiger could die out has come as a national shock.

At an international conference in New Delhi, Kamal Nath, India's environment minister, won agreement from 9 of the 14 "tiger range" states in Asia—countries where at least some tigers survive in the wild—to join in establishing an organization, the Global Tiger Forum, to coordinate measures to combat poaching and to preserve tiger habitats. But as the conference ended with ambiguous commitments from many of the countries that attended, and with none from China, which was among the countries that shunned the meeting, Mr. Nath echoed fears that tigers could soon disappear in the wild.

"If there are no new efforts made now," he said, "it will not take more than a decade to see the tiger go."

In the bid to save the tiger, India finds itself in an undeclared alliance with the United States, where President Clinton has lent his support to steps that could place United States trade sanctions behind the battle to stop the tiger poaching. At a meeting in Geneva, Interior Secretary Bruce Babbitt is expected to announce what action, if any, the United States plans to take after issuing warnings last fall to China and Taiwan, which have been identified as the principal culprits in an underground trade driven by the use of tiger parts in traditional Chinese medicine.

For centuries the tigers' main enemies were hunters, who coveted them as trophies, and later a fashion industry that made an expensive accessory of the tiger-skin coat. But with the banning of tiger hunting as a sport in India and most other tiger-range nations nearly a quarter of a century ago and the outlawing of the trade in tiger skins, the threat to the tigers has shifted to poachers who have targeted them for the parts hunters once left as carrion: the skull and bones, the whiskers, the sinews and the blood.

The trade is driven by booming markets for ancient Chinese medicines and potions made from tiger parts. In Hong Kong, China and Taiwan, and in Chinatowns across Europe and North America, Chinese apothecaries do a steady trade in tiger wines, tiger balms and tiger pills, celebrated among Chinese and other Asian peoples for their supposed powers to treat rheumatism, to restore failing energy and to enhance sexual prowess, as well as for the treatment of rat bites, typhoid fever and dysentery, among other ailments.

Among conservationists, Mr. Babbitt is seen as standing at a turning point not only for the tiger but in the wider battle for the survival of wildlife. Sam LaBudde, an Indiana-born biologist who has traveled widely in the Asian nations engaged in the tiger trade, gathering evidence for the Earth Island Institute, a conservation group based in San Francisco, sent a letter to Mr. Babbitt after attending the New Delhi conference, saying the interior secretary's stand in Geneva would be a bellwether for the conservation movement as a whole.

In the letter, Mr. LaBudde argued strongly for punitive actions against Taiwan, saying evidence he gathered on a visit there last month, including visits to 15 apothecaries in Taipei and three other cities where tiger-bone preparations were freely available, showed that Taiwan's compliance with United States demands was cosmetic.

"Beyond the very real question of whether tigers survive in the wild," Mr. LaBudde said, "the entire east Asian community is waiting to see whether the issue of species conservation is something that must be addressed substantively, or simply dismissed as a trifle."

The tiger's emergence as a priority species for wildlife conservationists was hastened by events in June 1992 at Ranthambhore, 200 miles south of New Delhi in the old princely state of Rajasthan. Gopal Mohgiya, a member of a tribe of hunters who traditionally worked as guards for local herdsmen, was arrested as a poacher by the police at Sawai Madhopur, the local railway junction, with the skin and bones of a freshly killed tiger in his possession.

The arrests of several other men followed, including a local Muslim butcher, Gulam Hussain, who was accused of rendering tigers poached from the reserve at Ranthambhore. According to police accounts, the men confessed to poaching at least 12 tigers, and selling them to middlemen who traded them to an underground market in tiger parts operating in the crowded bazaars of Old Delhi. Soon, similar reports of tigers lost to poaching began appearing from many of India's 19 other reserves.

The poaching, now estimated to have accounted for the deaths of as many as 1,600 tigers in the five years before the arrests at Ranthambhore, punctured Indian complacency. For 20 years, the government in New Delhi had basked in the approbation of conservationists for the success of Project Tiger, established in 1973 by Indira Gandhi, then the prime minister. By establishing and later expanding a network of tiger reserves, many of them in former hunting grounds like Ranthambhore, India claimed to have brought the tiger back from the brink of extinction.

According to an official tiger census, the figure of 1,800 tigers surviving in the wild in India in 1973 had grown to 4,300 tigers by 1992, an accomplishment with few parallels in the history of wildlife protection. But in the aftermath of the poaching scandals, Project Tiger was caught in a maelstrom of allegations. At best, its officials were said to have succumbed to bureaucratic instincts, inflating tiger counts to please politicians in New Delhi. At worst, they were denounced as crooks, conniving with poachers by turning a blind eye to illicit hunting in the reserves.

After a battle that went to India's Supreme Court, a new census was ordered, involving independent conservationists. Strict application of methods used for generations—plaster casts and tracings of tiger "pugmarks," or pawprints, found on forest trails, and sightings of tigers appearing at water holes—produced stark results. Ranthambhore, which had reported 45 tigers in 1991, was found to have only 28 tigers. Similar counts elsewhere, and guesswork about the two-thirds of India's tigers believed to live outside the reserves, led some of India's top tiger experts to conclude that the real count could be as low as 2,700.

Mr. Nath, the environment minister, announced a sweeping program to improve protection of the tigers, including the creation of an antipoaching strike force. But the strike force remains on paper only. Measures to improve primitive methods of agriculture and animal husbandry around reserves like Ranthambhore are in their infancy, where they exist at all. In Sariska, another tiger reserve in Rajasthan, 400 illegal marble quarries operate under the eyes of corrupt officials, devastating the forest cove and lowering the water table that keeps the surviving forest alive.

Indian wildlife experts who have conducted their own investigations say that poaching continues to thrive, driven by greed all along the illicit trading chain, from tribesmen stalking tigers by night in the reserves to apothecaries in Taipei. At the level of the village poacher, the rewards, perhaps $100 for a complete tiger, amount to no more than a normal year's income, perhaps more. By the time the bones, claws, eyes, penis and whiskers reach Taiwan, the value can increase several hundredfold.

One study, by Traffic, the wildlife monitoring arm of the World Wide Fund for Nature, showed that a tiger penis can sell for $1,700 in Taiwan, and powdered tiger bone for as much as $500 a gram. A single tiger can yield as much as 11 kilograms of bones, or more than 24 pounds. Investigators for Traffic worked in secrecy, sometimes posing as potential buyers of tiger parts, after several wardens in the tiger reserves were murdered on suspicion of betraying poachers, including two in Ranthambhore.

Ashok Kumar, Traffic's director in India, has given the Indian police the names of several men he believes are at the heart of the poaching network, operating from headquarters in Bombay and Delhi, but none of the leaders have been arrested. "Everything comes down to the city guys who are the brains behind the poaching, and we've never gone after them," Mr. Kumar said. "All the effort has gone into catching the poacher, the little guy, who earns a miserable three thousand rupees," equivalent to $100.

The trunk line Traffic traced in the tiger smuggling, by air and rail to the Himalayan foothill towns of Simla, Srinagar and Ladakh, and then by mule and yak across 15,000-foot passes into Nepal and China, is only rarely interrupted, and then mostly for small-scale seizures. Chinese officials won praise last year after holding a well-publicized bonfire of seized tiger bones, but Mr. LaBudde and other conservationists believe little has been done to inhibit the flow of the smuggled tiger parts across China into Hong Kong and Taiwan.

At Ranthambhore, meanwhile, the poachers arrested in 1992 have recanted their confessions, and are predicting acquittal if their case, already deferred for nearly two years, goes to trial. Mr. Hussain, the butcher, told a visitor that the only tigers killed were ones that strayed out of Ranthambhore and stalked villagers' cattle. "But don't misunderstand," the 65-year-old butcher said. "I am all for the tiger being saved. He is the king of the jungle, and once he is gone, the forest will go with him. Then we will have nothing."

—JOHN F. BURNS, March 1994

Hyenas' Hormone Flow Puts Females in Charge

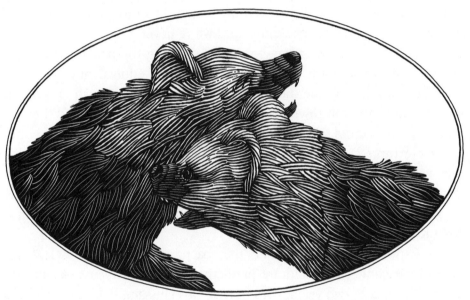

Brian Callanan

BY POPULAR REPUTATION, hyenas are repulsive scavengers, their coats a mass of mange, their mouths rabidly agape, their laughlike calls psychotic in pitch.

So it is a splendid surprise to discover, upon visiting a research colony of spotted hyenas sequestered in pens on the hills of Berkeley, that the beast is in fact a beauty, a creature every bit as baronial as that more celebrated carnivore, the lion. Its espresso-brown face is at once tender and strong, familiar and alien, combining a bit of bear, a hint of snow leopard, even a flicker of harbor seal. Its rear legs are considerably shorter than its front, a body plan that allows it to run extremely long distances in pursuit

106

of its prey, and its chest and neck are a mesh of dense muscles, powering a jaw that can easily crush the skull of a large antelope.

And pulverize prey the animal does: Far from feeding on leftover carrion, as lore would have it, the spotted hyena is the most ferocious of hunters, accounting for much of the game killed on the Serengeti Plain and in other regions of sub-Saharan Africa where it abounds. It is also the most efficient consumer, devouring flesh, bones, hoofs, teeth, fur—everything but a few bits of horn.

"In less than thirty minutes, a group of two dozen hyenas can reduce a five-hundred-pound adult zebra to a blood stain on the ground," said Dr. Laurence G. Frank of the University of California at Berkeley, an animal behaviorist who studies hyenas. "They eat so much bone that their feces look like chalk."

But what is the most outstanding feature of the spotted hyena, and the main reason why Dr. Frank, Dr. Stephen E. Glickman and other researchers have built a large facility to study the creature, is its bizarre balance of hormones.

While in the womb, male and female fetuses alike are exposed to extraordinarily high levels of male hormones, particularly testosterone. The hormones originate in the mother's ovaries and pass through the placenta, and they have a dramatic effect on the developing cubs. As a result of the androgen bath, both sexes end up with masculine-looking genitals, the male bearing the standard equipment, the female having an extremely enlarged clitoris that resembles a penis and fused, protuberant vaginal labia that look like a plump pair of testicles. Both sexes can and do get erections at the slightest provocation—when sizing up a stranger, when greeting a friend. But though the two sexes look equal, they are not: The female is in charge.

The researchers are exploring the hyena's unusual endocrine system to understand how the mother's reproductive machinery ends up generating such high doses of androgens that in turn profoundly influence her growing cubs. They believe their work will yield real insights into many puzzles of physiology and behavior, among them how androgens and their female counterpart, the estrogens, jointly influence sexual development in all mammals.

Already their discoveries are toppling traditional ideas about testosterone and its role in shaping dominant behavior, suggesting that for many animals, including humans, other hormonal pathways may be more important in the genesis and fine-tuning of a ferocious personality.

"This is a study of how males become males and females become females," said Dr. Glickman. "We're using a strange animal to test ideas that can then be applied to more conventional cases."

The work may also lead to a more sophisticated understanding of the link between hormones and aggression. Hyena cubs, when they emerge from their testosterone-laced uterus, are the most belligerent newborns among mammals, so wired for a fight that they immediately begin attacking one another, often to the death of one, particularly if the two cubs are the same sex. Among mammals, such sibling murder is extremely rare. But aggression in the hyena is not simply a matter of excessive male hormones. As they age, the levels of testosterone in female hyenas decline significantly below that of the males, and yet the females remain on average far more pugnacious and slightly stouter. Indeed, when Dr. Glickman tossed pieces of horse meat into one pen at the Berkeley facility that held a male and a female hyena, the male did not even bother scrambling for his share of the snacks until the female had her fill.

"At this point, he defers to her without giving it a second thought," Dr. Glickman said.

Despite their virilized anatomy and domineering behavior, female hyenas perform their feminine roles adroitly, managing to copulate through a tiny opening in the clitoris and then give birth through that same phalluslike organ—unique activities that are made easier by well-timed increases in estrogen to help the skin soften and stretch.

The scientists said their project is one of the few that are bringing together field research with detailed biochemical and molecular analysis of the hormones and genes that shape animal behavior. That convergence has attracted the attention of other specialists, including Dr. Paul Licht, who normally studies amphibians at Berkeley but was drawn into the hyena project to perform the endocrinology experiments, and Dr. Pentti K. Siiteri, now a professor emeritus of obstetrics and gynecology at the University of California, San Francisco, who saw the hyena as a magnificent opportunity to better understand the hormonal yin and yang, the estrogens and androgens, at work within humans.

Dr. Siiteri has evidence that a condition called polycystic ovarian syndrome, a common disorder in which women produce abnormally high amounts of androgens and are often infertile, may be caused by a hormonal event similar to one at work in pregnant female hyenas. "The hyenas are leading us to new insights on old problems," he said.

The researchers are also studying classic problems of animal social behavior, the rituals and transactions of everyday hyena life. Dr. Frank alternates between working at Berkeley, where he can observe the colony of 38 spotted hyenas in close and defined circumstances, and at the Masai Mara National Reserve in Kenya, where he has been following a clan of about 70 wild hyenas since the late 1970s. In that way, he can take clues gleaned from the captive group and apply them to the authentic and more complex backdrop of the great savannah.

He and others have shown that hyenas observe an almost ironclad hierarchy, in which a reigning female and her offspring hold such sway over the others that, when a clan commences tearing some herbivore apart, a strapping adult male will capitulate to the most bantam cub of the dominant female.

"The hierarchy is astonishingly stable," said Dr. Frank. "The great-grandchildren of the matriarch I saw in 1979 are themselves at the top now, and those descended from the hyenas at the bottom are still at the bottom."

Spotted hyenas are the largest and most common members of the hyena family, a group that seems doglike but in fact is more closely related to cats, and is closest to mongooses and civets. The spotted variety is the only hyena species in which the female has virilized genitals, and as a result the animal has long attracted attention. Writing in his memoirs, Ernest Hemingway, an avid big-game hunter but a poor naturalist, repeated the long-standing myth that the hyenas were hermaphrodites. By the 1960s, scientists were secure in their knowledge that the female only looks masculine, but the biochemical mechanism explaining how she got that way remained to be learned.

In the mid-1980s, the Berkeley researchers decided to ship some hyenas back from Africa to study their endocrinology and behavior in earnest. They constructed large indoor-outdoor pens on a reasonably deserted hillside several miles from the university campus, brought over 20 infant hyenas and reared them by hand. The hyenas, reaching their adult size of about 200 pounds, maintained their innate ferocity among themselves, but took affectionately to their surrogate parents.

"A lot of them think we're their mothers," said Dr. Frank. "They're real lap lovers." Nevertheless, it can be risky working with animals able to crush and consume entire zebra femurs. "All of us have scars," said Dr. Frank, "but most of them are well deserved."

Many of the original immigrants have since bred in captivity, and 11 cubs have been donated to zoos around the country. Through studying the animals during pregnancy, the researchers have learned some of the hyena's more outstanding features. In most mammals, a male fetus takes on its masculine form courtesy of its own budding testes, which release testosterone that then sculpts the rest of the genitals; lacking that potent androgen tweak, a female develops female genitals. The mother does have trace amounts of testosterone circulating in her blood because all mammals carry some amount of the opposite sex's hormones, but in most species the placenta acts as a barrier, converting maternal testosterone into a harmless form of estrogen that cannot reach the fetus and thus has no effect on sexual development.

The spotted hyena is very different on a number of counts. The researchers discovered that although adult females do not have unusual amounts of testosterone in their blood, they do have high concentrations of another common mammalian hormone, called androstenedione (an-dro-steen-DIE-own), which is produced by their ovaries. Endocrinologists have long dismissed this as a junk or inactive hormone, but the California researchers have found that it is an important precursor to either estrogen or testosterone, although when or how it is converted into these active hormones in most animals remains unclear.

In the hyena, the conversion occurs in the placenta. Rather than act as a shield against maternal hormones, the hyena's placenta takes the precursor androstenedione and transforms it into fiery doses of testosterone. The fetuses of both sexes end up exposed to levels of androgens far exceeding what a male fetus can generate on its own.

In addition, the hyena gestation period is unusually long for a mammal of its kind, lasting about 110 days, two weeks longer than that of the much larger lion. During the extended gestation, not only do females end up with masculinized genitals, but all fetuses have a chance to mature. They grow so large that they tear the mother's clitoris as they descend through her unusual birthing organ, and they emerge with eyes open, muscles coordinated and teeth already erupted through the gum, also unusual for a newborn mammal. The combination of exposure to testosterone and their mature weaponry is often deadly.

"Most neonates root around for their mother's teat," Dr. Glickman said. "Hyena newborns root around for the back of their sibling's neck."

The scientists attribute that infantile hostility almost solely to the action of testosterone, but afterward, the hormonal profile grows more complex. As cubs, males and females still carry high doses of androgen in the blood, but the females, in becoming socialized among their peers, engage in far more exuberant and rough play than do males.

The scientists suspect that hormones other than testosterone are at work in ensuring the female's dominance, possibly the precursor hormone, androstenedione, which could influence behavior by linking up with the appropriate hormone receptors in the female's brain. They believe that they are on track to understanding these other hormonal pathways, work that could overturn traditional and simplistic dogma about the centrality of testosterone in fostering aggression. And since other female mammals, particularly primates like humans, possess significant levels of androstenedione, the results could at least partly explain the relationship between biochemistry and aggression in some women.

They believe that many aspects of hyena physiology evolved to ensure aggressiveness in the female. While feeding on a fresh kill, hyenas spiral toward a frenzy, hardly stopping to take a breath between bloody mouthfuls. There is no cooperative feeding or sharing.

Scientists suggest that such violent feeding behavior fostered the evolution of aggressiveness among the females, who had to fight to ensure their cubs had a chance of getting their share.

What is more, the females form the social backbone of any hyena clan; aunts, sisters, mothers and daughters all live together, with only a limited number of males permitted to loiter about to father their offspring. Males born into the clan must disperse upon reaching adolescence, and females guard their territory against unwanted, interloping bachelors. Hence, the structure of hyena society must have favored hormonal conditions giving rise to a race of animal Amazons.

"There seems to be a genuine advantage to female aggressiveness," said Dr. Licht. "So the real question may be, If it's such a good idea for hyenas, why didn't it happen more often?"

—NATALIE ANGIER, September 1992

4

WOLVES AND COMPANY

Ant societies are founded on chemical signals, but sociality in mammals, though it also uses scent cues, depends on the degree of intelligence necessary to handle social interactions. The dog family is special for its intelligence and sociality, as exemplified in the wolf pack and the hunting bands of the African wild dog.

Among the most interesting of long-term biological experiments is the saga of wolves and moose on Isle Royale in Lake Superior. A herd of moose swam out to the island around 1900, thinking they had found a moose paradise of trees and no predators. In 1949 some wolves ran across an ice bridge to the island and believed they had found wolf heaven—a realm of moose and no humans.

Since then the wolves and moose have been in uneasy equilibrium, a natural experiment that has been closely observed since 1958.

Wolves in the rest of the United States are doing better than those on Isle Royale. Despite controversy, they have been successfully reintroduced into Yellowstone Park and other former stamping grounds. Wolves are favored by ecologists and tourists but viewed askance by sheep farmers. The fox, too, is making a comeback, reaping the rewards of any animal that learns to live in environments shaped by humans. Suburbia, the fox has learned, is hospitable territory if you don't mind the human neighbors.

Red Foxes Thriving in Suburban Woods

ONE WARM EVENING not long ago, two anglers were leaving the Connetquot River State Park Preserve on Long Island, after an almost troutless day of fishing in the secluded nature reserve, when they spotted unusual movement in high grass at the edge of the woods.

Just off a park road, perhaps 15 feet away, the pointy-eared, white-muzzled, sharp faces of three young red foxes popped up. They flinched not at all and regarded their human visitors intently with bright-eyed curiosity. Cute—very cute. The fishermen agreed that the foxes had saved the outing.

Foxes? On Long Island? Yes, says Gil Bergen, the park manager. Not only that, he said, their numbers seem to have increased of late, and the young ones have grown bold—"they're quite cheeky." Both the growing numbers of the foxes and their self-assurance in the presence of humans are signs of a remarkable ecological success story of global dimensions.

In an age when so many wild species are under threat, their populations dwindling and their futures insecure, the red fox is thriving like few other wild predators. In fact, biologists say, it has become the most widely distributed wild meat-eating mammal on Earth, thanks to an evolutionary heritage that has enabled it to adapt superbly to the presence and activities of people.

Along with similarly adaptable creatures like raccoons, white-tailed deer, blue jays, mallards, Canada geese and many others, the red fox is a creature of the future, a likely survivor in a natural world increasingly chopped up, manipulated and dominated by people. Many scientists fear that as these superadapters proliferate and spread, a larger number of more

Like a Cat

In many ways red foxes not only resemble but also behave like felines, especially when hunting. These are some of the similarities:

- About same size as cats.

- Have good agility and balance; can tiptoe along a narrow rail.

- Hunt alone. Slink on belly like cats when after a bird.

- Light-enhancing membranes at back of eyeballs.

- Pupils can close to vertical slits.

- Long whiskers on muzzles act as sensors; the main function is to quickly identify where to bite prey to kill it.

Michael Rothman

A Plan of Attack

The pounce of a red fox is much like that of a cat. A fox can leap with unerring accuracy as far as 17 feet over level ground and come down, paws first, surprising its prey.

1  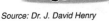 2  3  4

Source: Dr. J. David Henry

New York Times

sensitive and less adaptable species will be driven off the landscape, leading to a net loss of species and a relatively simplified, impoverished natural world.

To scientists who study the red fox, however, its success is good news indeed, not only because they consider it one of the most beautiful animals alive, but also because they have found it to be one of the most interesting and unusual. The behavior of the young foxes on Long Island suggested why: They focused their inquisitive attention on the human visitors through cats' eyes, with catlike intensity.

Field biologists have discovered that in many ways red foxes, unlike other members of the dog family, are as much cat as dog. They have the grace of cats, stalk prey like cats, slink like cats, hunt alone like cats, pounce like cats, have long whiskers like cats and are as agile as cats. (Unlike cats, however, they mate for life.)

On top of that, red foxes are literally as swift as the wind, having been clocked at 45 miles an hour over short stretches. They also run in a straight line, which is probably why they are the favorite prey of hunters on horseback.

Their signature move, called the "lunge and pounce," is a marvel of athleticism. They have been observed to leap as far as 17 feet over level ground in a single bound, from a crouch, to nab a mouse or rabbit unerringly. All in all, the fox's lithe build, russet coat, bushy white-tipped tail, black-tipped ears and legs and snowy white chest stir admiration among those who know it best.

"After all these years of studying the foxes, I am still really struck by their grace and their beauty," said Dr. J. David Henry, a Canadian government ecologist based in the Yukon, who has been closely observing foxes in the wild since the early 1980s.

"It's something I never habituate to," Dr. Henry said. "They are a constant source of inspiration to me, and it never wears thin."

Dr. Henry is the author of two books on foxes: *Red Fox: The Catlike Canine* (revised edition, Smithsonian Institution Press, 1996), and *How to Spot a Fox* (Chapters Publishing Ltd., 1993).

Not everyone sees the fox in positive terms, of course. It has long been associated in legend with sly deceit, and it is no favorite of farmers and homeowners whose chickens or pet cats have been preyed on by them, or owners of houses where a pair of foxes builds a den under a deck.

"Every night we have to lock in the chickens and pheasants," said Sophie Lowe, who with her husband, Tom, raises both kinds of birds not far from the Connetquot park. They rarely see the foxes except in the park after their young are born in the spring, she said. (Foxes breed once a year, carry their young for about 52 days and produce an average litter of five, whose members reach maturity in six months.)

Most people probably never know foxes are around, since they tend to be shy and secretive during the day and forage mostly from dusk to dawn. But they happen to love fragmented landscapes like farms and suburbs, which present lots of edge-of-the-woods habitat for the mice, voles, rabbits, and other small mammals that are the foxes' preferred prey. Experts educated in fox lore might detect them by listening for a high-pitched bark that sounds like a wail, or sometimes almost like a crow's caw, or a characteristic shriek, or any one of some 40 sounds foxes make. But they do not howl like wolves or coyotes.

"Although it's unsettling to many people to have foxes living near them, it might not be as unnatural as many people think," said Paul Rego, a Connecticut state wildlife biologist.

If the preferred prey is not available, foxes will eat almost anything they can subdue and swallow, and so can survive almost anywhere. The fox in the henhouse is an accurate image. So is that of the fox and the grapes; they have been known to subsist for long periods of time on a diet of 95 percent fruit. In suburbs, they forage in the garbage.

Some, abandoning their normal secretiveness, have become uncharacteristically bold, even in the daytime. In New Jersey's Island Beach State Park they have "become habituated to behaving like panhandlers," said Robert Lund, a wildlife research scientist with the State Division of Fish, Game and Wildlife.

In Connecticut, said Mr. Rego, there are frequent complaints about bold foxes loitering near homes. "They just lie out in a yard, chase the cats and don't run off when people are near," he said. "It's not uncommon at all."

Foxes can be a help as well as an annoyance. People who want to grab a gun when they see a fox "don't realize that foxes eat mice and rats, and they are more likely to cause a problem than the foxes," said Joe Ferdinandsen, a wildlife specialist with the New York State Department of Environmental Conservation, who is assigned to Long Island.

Some homeowners do prize the foxes. "I had a call last week from a lady who had a fox raising a litter near her house, underneath an outbuilding," Mr. Rego said, "and she said it was great."

Some people, Dr. Henry said, set food out for foxes. People who want to do that, he said, should make sure that the food is high in protein, and should not overdo it.

Suburbs present certain hazards for foxes, motor vehicles being a major one. Foxes killed on the road probably tend to be young juveniles making their first forays on their own, said Dr. Henry, adding, "If they survive their first summer and fall, I think, they're pretty traffic-wise."

Local populations of foxes rise and wane periodically, and although no studies have been made, wildlife biologists believe disease is a big factor in these fluctuations. A prime candidate is mange, which causes hair loss and can lead to death from infection or from winter cold. While foxes can contract and spread rabies, that does not seem to be a problem in the northeastern United States, where a different strain of rabies, carried by raccoons, is prevalent. Red foxes are not a primary carrier of that strain, Mr. Rego said.

Northeastern foxes in recent years have had to contend with a double threat to their hegemony: coyotes and the spread of forests. The bigger coyotes—which are as adaptable as foxes and have become abundant in the Northeast—can and do kill foxes, and the foxes respond by simply avoiding the coyotes' territory. The regrowth of northeastern forests has also reduced fox habitat by reclaiming the open farm fields that foxes like. On the other hand, suburban development is expanding fox habitat.

The foxes have more than held their own in the face of these conflicting factors. "Red foxes are just about everywhere," Mr. Lund said. In North America, Dr. Henry said, "they are much more prevalent than people realize."

The red fox today is found throughout most of the Northern Hemisphere and has been introduced to Australia and a few other places as well. Scientists consider all red foxes a single species, *Vulpes vulpes*. There were red foxes in North America before Europeans arrived, but relatively few. Some experts say the English settlers imported more from Europe, but Dr. Henry says the main reason for the fox's initial, phenomenal spread was the clearing of forests for agriculture, which opened millions of acres of prime fox habitat.

There are other kinds of foxes, of course, but for whatever combination of evolutionary imperatives, none are as bold or as successful as the reds. The American gray fox, for instance, is shier, more retiring and not nearly as widespread.

The red fox, Dr. Henry said, represents a superb evolutionary adaptation to its role as a specialized predator of small animals. Unlike bigger members of the dog family—the wolf, for instance—the fox does not gorge on one big, infrequent meal. Since its prey often comes in mouthfuls, it is a dainty eater rather than a ravenous gorger. This is because its stomach is extraordinarily small for the size of the animal—a trade-off in which the fox gains lightness and agility. It hides whatever it cannot eat for future consumption.

The fox has also evolved two catlike adaptations that aid in hunting at night: a light-enhancing membrane at the back of the eyeball, and pupils that close to vertical slits in the daytime to prevent blinding. But although the fox is about the size of some house cats (bushy tail excluded), its hind legs are proportionately much longer, the better to execute its patented lunge and pounce.

This maneuver is similar to one displayed by cats. In its short-range version, it looks as if the animal is merely frolicking as it crouches, leaps upward in an arcing motion, then comes down, front paws first. In longer jumps, the fox launches itself upward at a 45-degree angle and executes a leap that Dr. Henry has measured at 23 feet downhill as part of his observations in the wilds of northern Canada. "It's a gorgeous move," he said. "It's really something to see."

The fox uses the lunge and pounce when pursuing small mammals, he found. It cruises roadsides, trails and forest edges, looking and listening. When it hears a mouse, vole or rabbit in the grass or under the snow, it launches its leap, the purpose of which appears to be to carry the fox over grass and brush soundlessly, so as not to scare the prey away. A different hunting tactic is used when birds are the prey: The fox stalks the bird like a cat, slinking on its belly slowly, and then launching a low, catlike rush to nab the prey before it flies.

As with the cat, Dr. Henry believes, the fox's long whiskers have evolved as a means of guiding the killing bite once the prey is captured. The fox also has these sensory whiskers on the wrists of its front legs.

Sometimes, foxes go after bigger game. At the Connetquot park, Mr. Bergen observed a fox carrying a newborn deer that it had apparently killed. The body was still warm. On another occasion, Mr. Bergen said he saw four foxes—Dr. Henry believes they must have been juvenile litter mates—deliberately combing a field, in formation, apparently seeking fawns, whose protection strategy is to hunker down on the ground.

Once, Dr. Henry witnessed a particularly entertaining demonstration of the fox's speed as well as its guile. Two coyotes were coming too close to a fox den to suit the female (called a vixen) that lived there. Feigning injury, she tantalized the coyotes into chasing her. She would stay just out of reach, all the while moving away from the den. This tense nip-and-tuck drama went on for half an hour. Finally, when the vixen thought the game was won, she simply ran away from the slower coyotes.

"She put it into high gear and she was gone," Dr. Henry said. "The coyotes just stopped and watched her. She left them there with their mouths open."

—WILLIAM K. STEVENS, May 1998

Wolf's Howl Heralds Change for Old Haunts

Reaction at Top of the Food Chain
Ecologists expect that returning the wolf to its place as top predator in the Yellowstone ecosystem will transform relationships among many species of prey, predators and scavengers, and even change soil conditions and vegetation.

THE SERENE, SNOWY valley of the Lamar River in Yellowstone National Park would be a cafeteria for wolves at this time of year. Thousands of elk, deer and bison, driven from higher elevations by winter, congregate in the valley. But wolf packs have not prowled it for decades, so the animals that would be their natural prey live there in comparative peace.

Now, in an ambitious restoration project, two packs and the nucleus of another—14 wolves in all, the first of a group that could eventually grow to hundreds—have been living for several weeks in three one-acre pens within the valley and are soon to be set free. When they are released, the result is likely to be an ecological upheaval whose ripples will spread in time to the farthest, most obscure corners of a region the size of West Virginia. It is the largest nearly intact ecosystem remaining in the temperate latitudes of North America, and with the wolf's reappearance it will become more or less complete again.

The howl of the wolf haunted the Northern Hemisphere long before humans arrived. Wolves dominated those landscapes throughout most of history, occupying the summit of the ecological pyramid until recently, when they were shot, trapped and poisoned nearly out of existence. Perhaps half a million inhabited the world once, but now only about 100,000 to 150,000 remain, including some 2,200 in the United States south of Canada, all but about 200 of them in northern Minnesota. The species is officially listed as endangered in the contiguous 48 states except Minnesota, where it carries the less serious designation of threatened. But scientists have lately peeled away layers of myth and misunderstanding, and the resulting revolution in attitudes has opened the way for the wolf's return to some former homelands.

Of all the major elements that made the Yellowstone ecosystem function before white explorers visited the region, only the wolf is still missing. Now modern ecologists will have an uncommon chance to observe in detail the difference made by the presence or absence of a top predator like the wolf.

"It's an exciting opportunity for a biologist," said Edward E. Bangs, a wildlife biologist for the United States Fish and Wildlife Service based in Helena, Montana, who heads the northern Rockies wolf reintroduction project.

The Yellowstone venture is part of a regional effort over the next three to five years, also involving central Idaho, to rescue the species, *Canis lupus,* from its endangered status. The project has been beset by criticism and lawsuits from ranchers and their allies who fear that the predators will kill cattle and sheep. To assuage their fears, the government has agreed that any wolves proved to have killed livestock can also be killed, and the Defenders of Wildlife, a conservation group, has set up a fund from which ranchers are to be compensated for any losses. Still, the venture continues to draw the fire of some ranchers, and it was the subject of hearings last week in the Republican-controlled Congress.

One of 15 wolves released in central Idaho this month has just been shot and killed, and the killing involved an attack on livestock, according to local officials.

Assuming the Yellowstone part of the project goes forward on schedule, its wolves will be released from the pens in late February 1995 or early March. Wildlife biologists hope that by then a homing instinct that might propel them back to Canada, where they were captured earlier this month with the aid of nets and tranquilizer guns, will be broken.

"If they are released right now, I think they're all going to head north," said Dr. L. David Mech, a wildlife research biologist with the National Biological Service based at the University of Minnesota at St. Paul and a longtime wolf researcher who has served as an advisor to the reintroduction project. Already, he said, some of the wolves recently set free in Idaho have headed north.

Wolves have naturally reestablished themselves in Glacier National Park, bordering Canada 250 miles northwest of Yellowstone. While it is possible that they might eventually move south, Dr. Mech said it might take at least a decade for functioning packs to become established in Yellowstone. Two confirmed sightings of lone wolves have been made in the Yellowstone region, but no breeding populations are known to live there.

If the reintroduced wolves remain in Yellowstone, scientists believe, a transformation in relationships among the ecosystem's wild inhabitants will take place. Not only will there be occasional and sometimes deadly competition and combat between wolves and the region's other big predators—bears, cougars and coyotes—but changes will also be seen in the health of the elk, deer and bison populations and in the welfare of a host of scavengers. Scientists also expect to see changes in the vegetational profile and even the chemical makeup of the soil.

In a hypothetical example given by Dr. Mech, a wolf kills a moose. The remains slowly disintegrate and add minerals and humus to the soil, making the area more fertile. Lush vegetation grows, which attracts snowshoe hares, which in turn draw foxes and other small predators, which coincidentally eliminate many of the mice that live nearby. A weasel that used to hunt the mice moves to another area and in so doing is killed by an owl. The chain could be extended indefinitely.

Wolves also exert what biologists consider a pruning effect on their prey populations, zeroing in on young, weak, old and sick victims and providing a constant supply of carrion throughout the winter for scavengers like ravens and part-time scavengers like foxes and eagles. Otherwise, many weak prey animals survive the winter only to die in greater numbers than can be absorbed in the spring. The size of prey populations is somewhat controlled by wolves as well; Dr. Mech expects Yellowstone's burgeoning elk herds to be reduced by as much as 20 percent, enough to relieve heavy pressure on plants. This would change the mixture of plant types, since ungulates eat only certain species.

No animal has worked more powerful magic on the human imagination than the wolf. From werewolves to the story of Little Red Riding Hood, Western culture has demonized the species. But by studying wolves in both nature and captivity and by investigating reports of their past behavior, scientists have dashed most of the myths. No one has ever documented a fatal attack by a nonrabid wolf on a human in North America, and reported attacks elsewhere in the world have tended to fall apart on close scrutiny. Wolves, in fact, have been known to run away from humans, whining, while humans removed pups from the wolves' dens. "They're very afraid of people," said Dr. Mech.

Nor have any of the many reports of human children being raised by wolves, going back to Romulus and Remus, ever been substantiated. And wolves do not howl at the moon. Howling is a means by which competing packs warn each other of their presence and, within the pack, of bonding socially; group howling sometimes seems to be something like a community sing.

What Dr. Mech and others have learned about wolf reality is perhaps more fascinating than myth. Wolves, for instance, are as various in their personalities as dogs, their lineal descendants—and as humans. Their social life within the pack is a mixture of dominance and what people would call affection. The pack is a highly structured hierarchy with a dominant male and female, the alpha pair, at the top and close social bonding among the other members, which in large packs may include other breeding males and females. Usually, these other breeders leave to form their own packs.

Removed from the pack, a wolf pup, like a dog, will bond to a human. Wolves smile at one another, lick each other's faces, engage in tail-wagging play and sometimes snuggle. Dr. Mech has observed wolves lying close together and putting their paws around each other's shoulders as if hugging. But they can instantly jettison playful affection to become aggressive killers. They show no hesitation in dispatching a strange wolf that happens to intrude on their territory, not to mention strangers of other species, like coyotes.

The three initial Yellowstone groups, penned in by chain-link fences at sites separated by two or three miles, break down as one pack of six wolves, another of five, and a third, nuclear pack consisting of a female and her female pup plus a large, 120-pound alpha male who first met the other two in the Yellowstone pen. After an initial ritual of biting and posturing, according to Sheryl Matthews, a park spokeswoman, the three socialized quickly, with the adults displaying equal dominance—a potential alpha pair. All three are gray.

The largest pack consists of a gray alpha pair, two young black males, an adult gray female and a young gray male. (Pack sizes can range up to nearly 30, but average fewer than 10.) The alpha male of the second pack is described as an unusual bluish-colored wolf. His pack mates are two adult gray females, neither of which has yet emerged as the alpha, a young black male, and an older gray male. The mixture of colors denotes considerable genetic variability, an important consideration in reintroduction.

The time when the wolves are scheduled to be released would normally be breeding season. After a gestation period of about two months, pups would be born in April or May, and pack activity would revolve around the underground den in which the mother (most often there is only one in the pack) raises the pups. Other pack members range far and wide on the hunt, bringing meat back to the den in their stomachs and regurgitating it for the mother and the pups. After the pups leave the den, the pack sets up headquarters in an aboveground rendezvous area. When the pups are about six months old and able to travel, the pack goes on the move, ranging over a territory of up to hundreds of square miles (smaller where prey is more densely concentrated, as in Yellowstone) in a constant search for food.

The stress of transfer and captivity could preclude breeding by the Yellowstone wolves until next year, said Dr. Mech. It may also take them some time to figure out patterns of prey movement, and this could make a difficult hunting job harder. Wolves fail as hunters more often than they succeed. Their main prey species have evolved with them, and so have developed formidable defenses. Wolves appear to respect sharp hooves, horns and antlers so much that a moose, elk or deer that stands its ground will likely discourage the predators from attacking. And unless the wolves catch their prey early in a chase, an ungulate easily outruns them.

"They've got to work for their living, no question," said Dr. Mech.

Wolves also must deal with competition from other big predators. Coyotes tend to follow wolves and move in on their kills, for which the penalty is usually instant death. Both Dr. Mech and Mr. Bangs expect the coyote population of Yellowstone, a major existing threat to the region's sheep flocks, to be substantially reduced by conflict with wolves.

Studies in Alaska, Canada and Glacier National Park have found that when grizzly bears and wolves come into contact, the two come out about even, on average; bears usurp some wolf kills, for instance, but wolves kill some bear cubs.

The mountain lion, or cougar, appears to be the wolf's chief rival. Recent studies of wolf-lion interactions in Glacier National Park suggest that although cougars kill at least as many big prey animals as wolves and possibly more, wolves dominate cougars in direct encounters. Wolves routinely follow lion tracks to steal lion kills or confront them aggressively, and wolves have chased cougars up trees and also killed them, according to studies directed by Dr. Maurice Hornocker, an expert on cougars. He is the director of the Hornocker Wildlife Research Institute, a private organization affiliated with the University of Idaho in Moscow.

Cougars, once nearly wiped out by human exterminators, have reestablished themselves naturally throughout the former range of wolves in the West, including Yellowstone. What will happen once the two of them are together in a relatively small area hemmed in by human habitation? Dr. Mech, saying that the first line of defense for both species is mutual avoidance, does not expect many problems. Mr. Bangs believes wolves may displace lions from lower areas, where lions now hunt but wolves would call home, to higher, rockier country that is the lion's natural terrain.

But what happens when the prey species concentrate in lower elevations in winter? Will wolves displace lions into inhabited areas, where they might come into conflict with people and prey on livestock? There is some evidence of this displacement in the Glacier area, said Dr. Hornocker.

"I feel we have an opportunity here for the first time in a hundred years to document the interaction of these two big predators," he said. But, he said, he is frustrated because he has not been able to secure funds for the studies. "This is the payoff that we'll miss."

One thing seems clear: As elsewhere, wolf packs will sometimes find themselves in direct conflict with one another, with fatal results. While packs assiduously try to avoid one another's turf, territorial boundaries must sometimes shift with the prey. The prey are largely spread out through the Yellowstone area during most of the year, and a wolf pack's territory will then be at its most expansive—perhaps 100 to 200 square miles. What happens when the packs converge on winter prey concentrations like those in the Lamar Valley? The wolves may shrink their territories and clump them together, increasing the chance of fratricidal border disputes.

Or will they? That is part of the mystery waiting to be solved. "To see how they arrange themselves and how their numbers respond to an area like this is what's fascinating," said Dr. Mech.

—WILLIAM K. STEVENS, January 1995

As the Wolf Turns: A Saga of Yellowstone

EARLY ON A CRYSTALLINE June 1997 morning, as sunlight began to brighten the peaks, forests and green meadows of the northern Rockies, the wolves known as Yellowstone's Pack from Hell were already at work.

Confidently, almost nonchalantly, five of the six adult members of what is formally called the Druid Peak pack patrolled their territory, sniffing the ground and depositing urine scent marks along an open hillside studded with patches of Douglas fir, spruce and aspen. Only 300 yards from a well-traveled road, a big gray female headed the single file, with the dominant pack member strolling along third in line—a 122-pound gray wolf, his outstretched, rigid tail advertising his status as alpha male.

He is no ordinary animal. Barrel-chested, powerful and aggressive even for a wolf (one scientist likened him to a linebacker), he ripped the bars out of an aluminum cage in which federal scientists transported him from Canada last year. "Right away we knew he was different," said Dr. Douglas W. Smith, a wolf biologist for the National Park Service who directs the reintroduction project.

No. 38, as this formidable creature is named, is part of what is widely hailed as a spectacularly successful effort to reestablish wolves in Yellowstone National Park. He solidified his reputation soon after arriving here by leading the Druids in a territorial raid on a longer-established wolf family, the Crystal Creek pack. The Druids killed the opposing alpha male and all his young offspring, banished the bleeding adult survivors to another part of the park, and took over their den. They sealed the takeover with a group howl. Since then the pack has killed two more adult wolves, earning it nicknames like the Pack from Hell and the Dreadful Druids. "They are a bunch of hoodlums," Dr. Smith said.

But that is only one part of the continuing story of the Yellowstone wolves, a tale of family loyalty, clan warfare, and overall population expansion whose twists and turns will largely determine the future of this most charismatic of all endangered species in America.

In an enterprise that has restored the only missing piece to Yellowstone's renowned array of big mammal species, 41 wolves have been transplanted to the park from western Canada and northern Montana since January 1995, and they and their offspring have produced an estimated 60 to 70 pups. (The exact number will not be known until early winter, when this year's pups are nearly grown and a better count can be made.)

Since the program started, 10 pups and 10 adults have died from various causes.

Here, on an open and easily accessible stage, the day-to-day struggles that determine long-term evolutionary dominance are on unusually clear display. And in this arena it turns out that for all their aggression, the Druids have so far been eclipsed by another group, the Rose Creek pack, and especially by its central figure. She is an aging, long-suffering, charcoal-colored alpha female, called Rosie by some, Mom or Big Mama by others, but simply No. 9 (the number assigned to her when she was transported from Canada in 1995) by most.

No. 9 is the undisputed matriarch of Yellowstone wolfdom. With seven adult members, her pack is the largest of nine in and around the park. Only 30 months after the first wolves were transplanted here, there may be as many as 100 ranging freely hereabouts. If things keep going this way, federal officials say, it may be possible to take the wolves off the federal endangered list by 2001, ahead of the projected schedule. More than any other Yellowstone wolf, No. 9 is responsible for this success: Almost half the wolves here are her children or grandchildren.

This year alone, in a highly unusual feat of reproduction, her pack produced three litters of pups. An older daughter has formed a pack of her own, the first naturally formed such family grouping here since wolves were exterminated from Yellowstone by the government in 1926. And as No. 9's other descendants mature and form still more packs, her genetic legacy will spread farther.

No. 9 attracted widespread attention two years ago when her original Yellowstone mate, No. 10, was shot and killed, leaving her to fend for herself and eight pups, the firstborn of the restoration experiment. Now she not only has the upper hand in terms of reproductive and genetic dominance; she and her group have also shown their strength in battle, aggressively driving off the invading Druids. But there was a price. One of her yearling sons—"feeling his oats," Dr. Smith said—apparently pursued the fleeing Druids alone. They turned and killed him. And just a few weeks ago, the Druids caught one of No. 9's daughters alone and killed her, too. The daughter's pups, one of this year's three Rose Creek litters, died.

Now there seems to be a standoff in this evolutionary numbers game, albeit a tenuous one since the Rose Creek and Druid Peak territories abut each other. No. 9 has surmounted all this and more to emerge grayer, distinguished looking and a near legend. If anyone ever erects a statue of a wolf in Yellowstone, says Dr. Smith, "it should be her."

In all, an estimated 85 to 95 wolves now live here, and Dr. Smith says that given normal reproduction rates there could well turn out to be 100 when the count is complete. Officials of the northern Rockies wolf recovery program say that no more introductions are necessary to meet the goal of removing the wolf from the endangered list. Natural reproduction, they say, should do the rest.

"We could delist by December 31, 2000," said Edward E. Bangs, a biologist with the United States Fish and Wildlife Service based in Helena, Montana, who heads the northern Rockies recovery program. For that to happen, there must be at least 100 wolves and 10 breeding packs three years running in each of three areas: Yellowstone, central Idaho (where another population has been successfully introduced) and northwestern Montana (which has been naturally recolonized from neighboring Canada). There is a "fifty-fifty chance" that the three-year countdown to delisting will begin next year, Mr. Bangs said.

The eventual hope is that as the three wolf populations expand, they will begin interbreeding with one another, creating a single, large northern Rockies population with a large, healthy gene pool. Other imperiled American gray wolf populations, chiefly in the Great Lakes region and the Southwest, are being treated separately for delisting purposes. Wolves are plentiful and in no danger in Canada and Alaska.

The big surprise in Yellowstone this year has been the birth of multiple litters of pups to three packs: the three litters of No. 9's group and two each in the Druid Peak pack and a group called the Chief Joseph pack. Partly, some experts say, this may be an artifact of wolf reintroduction.

Usually, members of a pack would all be part of the same biological family, with only the alpha pair breeding, and inbreeding would not be a regular occurrence. But here, most packs have been constituted by throwing together unrelated wolves. The upshot is that since the alpha male is not related to the pack's females, he can mate with them all without causing social disruption.

But the primary factor in producing multiple litters, many biologists believe, is simply that Yellowstone abounds with prey, especially elk. The wolves are "living on a high plane of nutrition," said Mike Phillips, a wolf biologist who recently left the reintroduction project after directing it from the start. Dr. L. David Mech, another federal biologist and wolf expert, said that "you don't have to be a rocket scientist or even a wolf biologist to predict that with that amount of prey, once they get settled into the area, it's a matter of converting all that prey to new wolves, which they do well."

So rich are the pickings that the Druid Peak pack, for instance, last fall killed 21 elk in a month, 18 of them calves. Nor do the wolves have to search very far to find a meal. Most of the packs' territories are consequently small. The first naturally formed pack in the park, called the Leopold pack after the great conservationist Aldo Leopold, has what scientists believe may be the smallest and best wolf territory in North America: a relatively tiny turf about six miles by eight, just a short distance from park headquarters at Mammoth Hot Springs.

The easy availability of natural prey is also cited as a possible reason why there have been so few cases of Yellowstone wolves' killing livestock. In 30 months, only four instances of livestock depredation by the wolves have been reported, all involving sheep. The owners were compensated for their losses, a total of $2,186, from a special fund maintained by the Defenders of Wildlife.

But one group of wolves, the Soda Butte pack, was recaptured and transferred out of its territory on private land just north of the park where, Dr. Smith said, "they lived among cows and sheep and didn't touch a one." They were moved farther south into the park nevertheless, after threats were made on their lives. Tourists have also sometimes interfered with the wolves. In at least two cases, Dr. Smith said, overenthusiastic visitors have in effect chased wolves away from a kill by getting too close, an illegal act in Yellowstone.

For the most part, however, the wolves seem to be securely established and thriving, and Dr. Smith said that a second, more scientific phase of the project was now beginning in which their behavior was to be studied over the long term. A platoon of volunteers is staking out the packs in relays, tracking their movements with radio direction finders every half hour for 48 hours at a stretch (several wolves in each pack wear radio collars) and trying to monitor their behavior through binoculars and telescopes.

It is in the area of wolf-to-wolf relationships, both within packs and among them, that some of the most revealing observations have been made.

The life of No. 9, for example, reads like a soap opera. When she arrived here with a daughter from Canada in January 1995, she had been separated from her mate by trappers who captured her for a reward. A previous mate may have been killed. Here, she immediately mated with a big gray male, designated No. 10. He was about as strapping as No. 38, and together he and No. 9 produced eight pups. The Canadian daughter went off on her own and began foraging as a lone wolf. About the time No. 9's first Yellowstone pups were born on bare ground near Red Lodge, Montana, outside the park, No. 10 was shot by an unemployed carpenter. The shooter was later convicted of killing a member of an endangered species, assessed a $10,000 fine and sentenced to six months in jail.

Biologists rescued No. 9 and her pups and nurtured them in a pen until the pups were old enough to leave. Some of the pups escaped from the pen but stayed close, and a yearling male from the Crystal Creek pack, named No. 8, began hanging around and brought food to them. On her release from the pen, No. 9 immediately took up with No. 8, making him alpha male.

"Yearlings don't get promoted to alpha male all that often," Mr. Phillips said. "If she hadn't found his presence acceptable, she'd have run him out of there; but here's a young guy who knows the area, and she's got eight mouths to feed, and he seemed to like the pups."

No. 9's daughter from Canada also mated with a Crystal Creek yearling to form the Leopold pack, making her mother her sister-in-law as well. "This is truly a family affair," Mr. Phillips said.

Within the Rose Creek pack, a characteristic picture of autocracy moderated by cooperation has emerged. Observers say that one of No. 9's pups from 1995 appears to have emerged as the beneficent uncle; this year's pups overwhelm him with affection. He and No. 8, the alpha male, tirelessly carried food to the mothers and pups of the pack's multiple litters this spring.

Once, in a common display of the brutal nature of life in the wolf world, No. 8 sniffed out and killed an elk calf. One of last year's litter of pups distracted the mother elk while No. 8 began to feed. But a wolf pack is not a democracy: No. 8 bared his fangs at the yearling and would not allow him at the table. In a wolf pack, subordinates must mind their manners and bide their time until they can go off and form their own packs.

The next chapter of the story depends not least on how the simmering confrontation between the Rose Creekers and the Druids turns out. Wolf warfare is simple, Dr. Smith said: "Whoever's got more wolves wins." When the Druids savaged the Crystal Creek pack, they had a five-to-three advantage in fighting adults. The Rose Creekers had an eight-to-five edge when they repelled the Druids.

When the Druids "add their bevy of pups to their war machine," Dr. Smith said, "they're going to be a pack to contend with."

—WILLIAM K. STEVENS, July 1997

The Endangered Timber Wolf Makes a Surprising Comeback

THE HOWL OF THE EASTERN timber wolf can be heard these days in the far northern suburbs of Minneapolis and St. Paul, evidence that the species is making a surprisingly swift comeback around the northern Great Lakes.

In fact, the species could be removed from the endangered list ahead of schedule, because it is doing so well in Minnesota, Wisconsin and the Upper Peninsula of Michigan. The wolf's recovery has been accomplished without a reintroduction program, occurring largely through protection and public education.

"Wolves have turned out to be much more adaptable than we thought if they're left alone and have plenty of natural prey," said Ron Refsnider, a biologist at the United States Fish and Wildlife Service's regional office in Minneapolis.

Winter surveys of wolf populations are under way by plane and snowmobile in the forests of northern Wisconsin and the Upper Peninsula of Michigan, where the timber wolf had once disappeared as a breeding species. A count will be announced in the spring, but scientists say last summer's pup production probably increased the wolf population in the two states to at least 200.

"The Yellowstone wolf reintroduction is small potatoes compared to what's happening here," said Jim Hammill, a wildlife biologist with the Michigan Department of Natural Resources in Crystal Falls. "Wolves have returned in impressive numbers, and they've done it on their own, at no cost."

Late last winter, there were 80 wolves in 12 packs in northern Michigan, up from 57 in 1994. Wisconsin's wolves numbered 86 in 18 packs, an increase of 30 from the previous survey, and all of those packs raised pups this year.

In Minnesota, the wolf population has almost doubled since the first comprehensive survey, in the winter of 1978–79, counted 1,235 animals in the state's northeast corner, where there are large areas of roadless wilderness. Their present number is estimated at 2,200. Minnesota's wolves have expanded their range southward and westward to cover 35,000 square miles, nearly half of the state.

Not only are wolf howls heard in the suburbs of Minneapolis and St. Paul, said Bill Berg, a predator specialist with the Minnesota Department of Natural Resources in Grand Rapids, but wolves have moved from their forest haunts onto the state's western prairies, where they once hunted herds of bison. "Wolves continue to fool us," he said. "Ten years ago, I never would have predicted that we would have wolves near the Twin Cities. They've come to live closer to humans than anyone thought possible." Wolves, scientists say, pose no threat to humans.

The eastern timber wolf, and wolves in the Great Lakes area, are generally smaller than their northern Rocky Mountain counterparts. "Our males weigh 75 to 80 pounds, the females 60 to 65 pounds," said Adrian Wydeven, an ecologist with the Wisconsin Department of Natural Resources. "The wolves that were released in Yellowstone National Park weighed about 95 pounds." A wolf's size, he said, is governed by its main prey. "Wolves that hunt large ungulates like elk and moose tend to be larger, but the packs in Wisconsin, Michigan and Minnesota mainly hunt white-tailed deer and beaver."

Bounties were paid for killing wolves from 1865 to 1957 in Wisconsin and from 1839 to 1960 in Michigan, where they disappeared from the Lower Peninsula before 1900. When those states finally passed laws protecting the timber wolf, there were too few left to sustain viable populations. "The wolf was considered extinct in Wisconsin by 1960," said Mr. Wydeven. In Michigan, the last confirmed litter of wolf pups was born in 1954 and the surviving adults dwindled in the 1960s.

Minnesota stopped paying bounties in 1965, but the wolf was left unprotected until 1970, when Superior National Forest, the predator's stronghold, was closed to the taking of wolves by the federal government. In 1974, the timber wolf became one of the first animals to be listed as endangered under the new Endangered Species Act. The Minnesota wolf population was reclassified as threatened in 1978. And a revised recovery plan, published in 1992, recommended that the subspecies be removed from the list as soon as a second population of at least 100 wolves was maintained in Michigan and Wisconsin for five years.

Four years ago, wolf biologists estimated that the "date of recovery" would be the year 2005, and the speed at which wolves are reoccupying their former range has astonished the experts. "No one asks us anymore why we need wolves," said Mr. Refsnider. "Now, some people wonder why we need so many wolves."

Wolves returned to Wisconsin, emigrating from neighboring north-eastern Minnesota, in the mid-1970s, and state biologists began monitor-ing them in 1979. "We counted as many as 26 wolves one winter, but they declined to 15 in 1985," said Mr. Wydeven. "Since then, we've seen a steady increase every year but one."

Deer herds in the Great Lakes region, he noted, are at record high numbers and beaver are abundant because there are few trappers anymore. "An adult wolf will eat 18 to 20 deer a year, and the wolf is one of the few predators that can kill a beaver," he said. "So wolves have a tremendous prey base and there is a lot of unoccupied habitat." Mr. Wydeven said it would take another 15 to 20 years before wolves would saturate their habi-tat in Wisconsin.

A state forest near Black River Falls in west-central Wisconsin is the home of the southernmost breeding wolf pack in North America, Mr. Wydeven said, and one den is only a few hundred yards from Interstate 94. A "dispersing yearling female" that had been radio-collared 350 miles away near Ely, Minnesota, was struck by a car near Madison, Wisconsin's capital, he added.

It was not until 1989 that Michigan biologists confirmed the presence of a wolf pack in the western Upper Peninsula, near the Wisconsin border, and the first wolf pups in the state in more than 40 years were seen in 1991. Moreover, wolves crossing the frozen St. Mary's River from Canada have reoccupied the eastern end of the peninsula near St. Ignace as well as nearby islands, Mr. Hammill said.

"There is a lot of publicly owned forest land in northern Michigan that is still prime wolf habitat," he said. "Based on studies of satellite im-agery, we estimate our carrying capacity is around one thousand wolves. A few years ago, we wondered if the wolf would survive in the lower forty-eight states. The question now is, What are we going to do with this quickly growing population?"

Mr. Hammill and other scientists credit a change in the public's attitude toward large predators for the eastern timber wolf's fast recovery. In 1974, Mr. Hammill noted, four wolves from Minnesota were released near Marquette, Michigan, and all of them were illegally killed by hunters or trappers within a few months. "So much hell was raised that the biologists threw up their hands in surrender," he said. "People in Michigan weren't ready for wolves." Today, mange rather than illegal shooting is the main cause of wolf mortality, he said.

Mr. Berg said, "People have come to accept the wolf as a critter they can live with." One reason, he emphasized, is that wolves preying on livestock and poultry are promptly removed by animal damage control agents from the United States Department of Agriculture and farmers are fully reimbursed for their losses. In 1994, he said, 174 wolves were killed by federal agents in the state.

Mr. Berg also cited the timber wolf's adaptability. "When I first studied wolves twenty-five years ago, it was generally held that a pack would not cross a highway into new territory," he said. "Now that the pristine wilderness has all the wolves it can hold, they've moved out of the northern forest and we have new generations of wolves that do not consider either roads or people to be a threat. Minnesota now has the highest density of wolves on the North American continent, including Alaska."

—LES LINE, December 1995

Wolves May Reintroduce Themselves to East

AS THE ENDANGERED gray wolf makes a comeback in the Rocky Mountains and Upper Midwest, some conservationists are training their eyes on the Adirondacks as the next target for reintroducing the great predator. But the wolves may be way ahead of them.

Scientists say that wild Canadian wolves, following an ancient territorial imperative, are already expanding their range in Quebec toward the United States. A new study commissioned by the Wildlife Conservation Society, which has its headquarters at the Bronx Zoo and was formerly called the New York Zoological Society, places them within 40 miles of the Maine border. At least one lone wolf, and maybe more, has already entered the state.

The findings throw a new spotlight on a host of scientific and political complexities and difficulties involving wolf recovery in the Northeast.

In Maine, the study found, more space and less human activity make the prospects for the long-term reestablishment of the eastern timber wolf, a subspecies of gray wolf, more promising than in northern New York State's Adirondacks. Moreover, the study says, the Adirondacks are so isolated from the wolf populations living in Canada that natural recolonization is unlikely there.

In any event, some conservationists and scientists see natural recolonization as cheaper and less likely to stir opposition than capturing packs and flying them in from somewhere else, as was done recently in Yellowstone National Park. More than 50 wolves in 10 packs now live there.

But natural recovery may not be as simple as it looks. Some scientists say, for instance, that the wolf gene pool might be seriously diluted by the Northeast's ubiquitous coyotes. With the extirpation of wolves throughout most of the United States a century ago, coyotes have appeared in large numbers in the Northeast after expanding their range from the Southwest.

Left to themselves, wolf populations tend to expand until they inhabit all suitable territory available. Expansion often occurs when one member of a pack strikes out alone; a wolf can travel as much as 500 miles in search of a mate.

In southeastern Canada, where the eastern timber wolf is abundant, geneticists have found evidence that interbreeding between wolves and coyotes has been common. If the same thing were to happen in the northeastern United States, wolves might eventually become so hybridized that their identity would be blurred. The identity of the southern red wolf has been clouded by just such interbreeding. Some experts consider it a bona fide subspecies of wolf, while others say it is a wolf-coyote hybrid.

Some scientists say hybridization might be avoided in the Northeast by following the Yellowstone model and artificially transplanting packs of "pure" timber wolves from northern Quebec, where there are not yet any coyotes. Presumably, as has happened in the Rockies, the ready availability of wolf mates, a cohesive pack structure and the wolves' sheer numbers would lead members of the packs to ignore coyotes as mates and to kill them or drive them away instead.

"The Yellowstone wolves have killed dozens of coyotes," said Dr. Robert K. Wayne, an evolutionary biologist at the University of California at Los Angeles who has studied wolf and coyote genetics. By contrast, he said, "our data suggest that wolves mate with coyotes where wolves are few and coyotes are abundant." If people want to restore wolves to the Northeast, he said, "they shouldn't piddle around with a few, they should do it in earnest" by bringing in four or five intact packs.

The developing debate is the latest chapter in an American recovery story that began when the gray wolf became the first animal to be listed as endangered under the federal Endangered Species Act of 1973. Although thousands of wolves still live in Canada and Alaska, preying largely on deer, elk and moose, the federal law mandates that wolves be protected in the contiguous 48 states, where they were long ago exterminated in most of their historic range by farmers, ranchers, bounty hunters and government agents.

Some 2,000 wolves in northern Minnesota were assigned the less serious status of threatened, and that population, along with some Canadian wolves, has now expanded into Wisconsin and Upper Michigan. About 100 wolves now live in each of those two states, and their gradual arrival has prompted relatively little outcry. Wolves also expanded naturally into Montana's Glacier National Park in the 1980s, where there are now about 100 in about 10 packs. In an attempt to speed the reestablishment of wolves throughout the northern Rockies, wolves from western Canada were introduced to Yellowstone and central Idaho in early 1995.

But the eastern timber wolf remains absent from the United States. That subspecies originally included both eastern and midwestern wolves, but taxonomists now consider the midwestern wolves part of the Rocky Mountain subspecies instead, and federal scientists have accepted the revision. That leaves the timber wolves of eastern Canada as the sole members of their subspecies.

The debate is only beginning on whether and how to reestablish timber wolves in the Northeast. The Defenders of Wildlife, a Washington-based conservation organization that spearheaded the Yellowstone reintroduction, wants to see wolves reintroduced into the Adirondacks. A study by the group last year found that the region had enough habitat and prey animals, like deer and beaver, to support roughly 150 wolves.

The more recent study by the Wildlife Conservation Society set out, as the first step in a long-term analysis, to assess the potential for natural recolonization in the Northeast. Two wildlife ecologists at the University of Maine at Orono, Dr. Daniel J. Harrison and Theodore C. Chapin, examined both the Adirondacks and Maine's North Woods, the two primary areas designated by the United States Fish and Wildlife Service as potential territories where wolves might be reestablished.

Based on the habitat criteria set by the agency regarding road and population densities, and on wolf densities observed in the Midwest, the researchers found that the Adirondacks might indeed support 140 to 600 wolves. But they also found that there was only 56 percent of the habitat necessary for long-term persistence of a wolf population isolated from other wolf populations.

And the Adirondacks are clearly cut off from Canadian wolves, the scientists found, leading Dr. Harrison to put the region's potential as wolf habitat "in the marginal category—it's not a sure bet." The study concluded that reintroducing wolves there would risk failure in the end. The finding that natural recolonization is unlikely in the Adirondacks also suggests that any reintroduction would require transplantation, as in Yellowstone.

Maine's 17,000 square miles of good wolf habitat, compared with fewer than 5,600 square miles in the Adirondacks, make it a better bet, the researchers found. "If I had to choose an area in terms of the probability of successful reestablishment," Dr. Harrison said, "I would choose Maine."

Joan Moody, a spokeswoman for the Defenders of Wildlife, said of the new study, "We're glad that it shows there is adequate habitat for wolves in Maine." But she said that reintroducing wolves there would not preclude the Adirondacks as a potential site. "They're not mutually exclusive," she said. "Our study shows there is wonderful habitat for wolves in the Adirondacks."

How might Quebecois wolves get to Maine? Dr. Harrison and Mr. Chapin identified two potential north-south strips of habitat that wolves could use to cross the St. Lawrence River Valley, a formidable barrier. One corridor is just upstream from Quebec City, in an area where intermittent ice cover may offer crossing opportunities. At the second corridor, just downstream from Quebec City, a large island narrows the river on both sides, and the home range of several wolves is just to the north.

Coyotes have been seen on the island, but Dr. Harrison noted that the St. Lawrence shipping channel and the four-lane highways that parallel it could prevent the successful dispersal of a significant number of wolves. Also, it is legal to trap and shoot wolves in Quebec. Whether these difficulties are prohibitive requires further investigation, Dr. Harrison said.

In 1993 and again last year, two large wolflike animals were killed in northwestern Maine. The first was identified as a female wolf, but genetic testing has not yet been completed on the second. At 81½ pounds, it was the size of a wolf.

Meanwhile, the Maine Department of Inland Fisheries and Wildlife has two trackers out scouring northwestern Maine in a needle-in-the-haystack search for wolf signs. It is "only prudent" to find and identify any migrating wolves before they suffer the fate of the first two animals, said Michael Amaral, an endangered-species biologist at the United States Fish and Wildlife Service in Concord, New Hampshire, whose agency is helping to finance the search.

If the two animals were indeed the vanguard of a wolf migration into Maine, Mr. Amaral said, "it's going to be a lot easier to allow those wolves to become established than to convince the ten to fifteen percent of the population that will never get on the side of the wolf to agree that the expenditure of millions of dollars of state and federal money is worthwhile."

—WILLIAM K. STEVENS, March 1997

The Most Social of Wild Canids Struggles for Survival

THE AFRICAN WILD DOG, one of the rarest and most quirkily beautiful mammals in the world, has more aliases than a member of the FBI witness protection program, and all of them are confusing. Take the name African wild dog: It sounds like an abandoned pet gone bad. A related name, "African hunting dog," is hardly better. The creature is indeed a superb hunter, but here, too, the phrase suffers from its human connotation, the whiff of English hounds out on an afternoon fox chase.

Finally, there is "painted wolf," the least accurate if most poetic tag of the lot. True, the dog's coat is a motley canvas of black, white and mustard, a furred version of combat fatigues. But the animal bears only a distant relationship to the wolf, having branched off from wolves, domestic dogs and other canids at least 3 million years ago.

Moreover, the African wild dog is much less aggressive than a wolf, and there is no such thing as the wild-dog equivalent of a lone wolf. To the contrary, the dogs are the most zealously social of all canids, perhaps of all mammals. Many social mammals can, in a pinch, survive and breed on their own, but the dogs are considered an obligate social species. They must live in a pack, and everything about their anatomy and behavior, from their radar-dish ears and fluttery, complex language to their floridly musky odor, underscores that need.

The name confusion is partly a result of the dog's obscurity. Until recently, nobody paid enough attention to make sure the animal had a consistent handle apart from its flair-free scientific designation, *Lycaon pictus*. Few researchers studied the dogs, and few zoos bothered owning any. In this country, only 18 of the 177 accredited zoos have African wild dogs.

"I don't know how many times people have come up to me after a talk and said, 'Gee, I never realized they were a distinct species,'" said Dr. Scott Creel. "Even biologists sometimes think they're domestic dogs gone feral."

Dr. Creel and his wife, Nancy Marusha Creel, who are affiliated with Rockefeller University, study the wild dog on the Selous Game Reserve in Tanzania. They visited the Philadelphia Zoo in November to check out the three pairs of wild dogs that recently arrived there from southern Africa, and to discuss with a reporter the plight and glory of a species that might rightfully be called an underdog. On their home continent, the dogs have lost staggering amounts of habitat and often have been shot as vermin. Once they roamed throughout sub-Saharan Africa. Today they are found in just eight countries and number a mere 4,000 to 5,000, making the dog second in endangered status only to the rhinoceros. By comparison, hundreds of thousands of the celebrated African elephant remain.

Now, however, it looks as though *Lycaon*'s fairy dogmother has finally arrived. Conservation groups have placed the wild dog at the top of their lists for attention. The Philadelphia Zoo, for example, recently designated the dog its flagship species and supports a number of dog study projects. Biologists are working mightily to gather basic information about the dogs, including how many live where, and why the creatures are so rare while other carnivores, like lions and hyenas, are so abundant.

The Creels are exploring the dogs' extraordinary social system, a type of cooperative breeding system in which only one pair of dogs reproduces in a pack, and the rest of the adults relinquish parenthood to serve as baby-sitters, nannies and even wet nurses for the alpha pair's pups. In Zimbabwe, Dr. Kim McCreery and her husband, Dr. Bob Robbins, are compiling a comprehensive vocal ethogram of the wild dog—a description of the many sounds the dogs make and how the dialects differ from one population to another.

The dog is also becoming famed among goggle-eyed safari-goers. Lions and tigers and hippos? Ho-hum. Where are the wild dogs? "Recently, a tourist complained for the first time to the park wardens here, because he was promised he'd see wild dogs, and he hadn't seen any," said Dr. McCreery, speaking by telephone from Hwange National Park. "This was wonderful news to us. It meant people are finally learning enough about the dogs that they really want to see them." And when a species becomes a tourism draw, she added, "there's hope."

By grace of the newfound attention and its protected status, and because it has no horns or tusks to beckon poachers, the dog appears to be holding its own in parts of Tanzania, Botswana, Kenya and elsewhere.

That tourists ever see the dogs at all is a modest miracle. One reason the dogs have not been studied much is that they are hard to find, and harder still to keep track of. A wild dog is relatively small for a creature that lives on big game. An adult weighs about 45 pounds, half the size of a spotted hyena and a seventh that of a female lion. Its blotchy coat serves as excellent camouflage when it is slinking through the woods. Its sounds are soft, easy to miss and surprisingly biodiverse. The African wild dog can meow like a cat and twitter like a bird, and though it whimpers and growls like a dog, it does not howl like a wolf or otherwise call attention to itself. In fact, said Dr. Robbins, many of its noises are above the range audible to the human ear.

Moreover, the dogs roam over enormous amounts of terrain, making it hard to keep up with a pack over time. A wild dog's home range averages 165 square miles, compared with about 10 square miles for a hyena. The only way to study wild dogs is to locate a pack and outfit one or two of the members with radio collars. "When we got started in Selous, we did nothing all day, every day, except try to radio-collar dogs," Dr. Creel said. "And still it took us five months to collar our first one."

As the Creels and other researchers have found, the key to understanding the African wild dog is relentless competitive pressure from other like-menued carnivores. Wherever they coexist, lions, hyenas, wild dogs and cheetahs all compete for sizable prey like wildebeest, impala, zebra and Thomson's gazelles. The predators each take a different tack to getting their share. Lions are huge and fierce and very rarely challenged. Hyenas are also powerful and threatening and have no compunction about stealing the prey of others. Cheetahs hunt during the day, when their rivals are asleep. And wild dogs, which live in packs of 4 to 20 adults and their dependent young, survive by raising teamwork to an art form.

Watching African wild dogs on the move is like watching a dance by Twyla Tharp, a gorgeous exercise in group kinetics. Even when confined to a zoo, if one dog gets up and bounds around, the others leap to their feet and follow suit. In the field, they make their hunting decisions through mass rallies, nudging one another excitedly and emitting birdlike calls. While trotting along in search of quarry, they often assume a parallel formation so that from the side it looks like one dog running. Their motions appear effortless, their slender legs barely grazing the ground as they glide forward at 10 to 30 miles an hour. On encountering a prey animal that looks vulnerable, they call to one another for support and move in jointly for the kill, with one grabbing the poor beast by the nose to subdue it, and the rest disemboweling it from behind. They usually hunt once a day, and they are so good at it that they complete about half the chases they begin. By contrast, lions successfully make a kill only about 10 percent of the time.

Yet sociality does more than make hunting easier. Because their predatory skills are known, the dogs are often trailed on a hunt by freeloading hyenas and lions, which seek to snatch away the prey once it is felled. A dog on its own is helpless against thieves. But assemble a team of the highly cooperative canids, and they have a shot at beating off even a gang of the bigger but less disciplined hyenas.

Most of the time, however, the dogs prefer to avoid encounters with competitors in the first place. For that reason, said Dr. Robbins, they have evolved inconspicuous means of staying in touch with one another, like their quiet calls. And to hear those calls, the dogs possess unusually huge ears, which look more like the ears of a prey species than those of a predator. The dogs also give off a pungent odor, again, scientists propose, as a way of maintaining pack contact.

Because of their interdependency, pack members usually treat one another with respect. "Aggression is muted in the wild dog," Dr. McCreery said. "They hardly fight at all, and they don't even bare their teeth at each other." Before going out for a hunt, they seem to count heads, she said, as though making sure the group is safely united.

For most species, a broken leg or other serious injury is a death sentence. But Nancy Creel said scientists had observed many cases where dogs would bring back food to a wounded mate, sometimes for months at a stretch. Adults also cooperate to rear young. All pack members will regurgitate chewed meat to the pups whenever they give a begging whimper.

Yet while group living has its advantages, there are costs as well. During the annual mating season, the ordinarily peaceable dogs turn snappish and hierarchical, with the alpha male battling to keep contending suitors from the alpha female. For their part, the subordinate females almost never bear young, and on the rare occasion when they do, their litters do not survive. Most astonishingly, the nonbreeding females appear able to lactate and nurse the alpha pups, an extraordinary investment for a female to make in any young but her own.

Why the subordinate females usually do not go into heat and conceive remains a mystery. In a recent study, the Creels and their colleagues could find no difference in stress hormone levels between subordinate and superior, dashing the notion that the senior animals maintain control through intimidation and fear. Perhaps the underlings do not mind biding their time. The Creels have observed that many on the bottom eventually get their chance at royalty.

The details of a dog's life could explain why the species is naturally uncommon, regardless of human affairs. With only one female per pack giving birth each year, the reproductive rate is not high. In addition, the desire to avoid lions and hyenas, while still having access to prey, could be why wild dogs demand such big home ranges. Right now, the dogs do best outside protected areas like the Serengeti National Park in Tanzania, where lions and hyenas abound, which means that the dogs' long-term conservation depends on something more than maintaining a few giant outdoor zoos.

—NATALIE ANGIER, December 1996

In Long-Running Wolf-Moose Drama, Wolves Recover from Disaster

THIS WINTER WAS LONG, tough and tumultuous on Isle Royale National Park in Lake Superior, scene of the world's best-studied predator-prey relationship, and the news is what wolf watchers were hoping to hear. A troubled wolf population has rebounded sharply and at the same time moose numbers have crashed in spectacular fashion.

Studies of wolves and moose and how their populations are linked began on the island in 1958 under the direction of Dr. Durward L. Allen, a wildlife ecologist at Purdue University in Indiana. The result is said to be the longest-running study of a wildlife population in the world.

For several years Dr. Rolf O. Peterson, who has been studying the park's wolf packs since 1970, has worried that the wolves might be on a fast track to extinction. Dr. Peterson saw their numbers crash from 50 in 1980 to 14 two years later, apparently from an outbreak of canine parvovirus, a deadly new disease that had ravaged dogs on the mainland and was brought to the island on hikers' boots. He watched the population stagger along at a dozen or so animals for more than a decade and grow top-heavy with old wolves that had little success at replacing themselves.

By 1993 there were only three females, all of them old, among the 13 surviving wolves. Just one of them had ever succeeded in rearing pups, and biologists feared that inbreeding had caused the isolated wolf population to stagnate.

Dr. Peterson, a wildlife ecologist at Michigan Technological University in Houghton, also watched moose numbers on the 210-square-mile wilderness island mount in the absence of the normally heavy wolf predation. By the winter of 1994–95, the Isle Royale moose population stood at 2,400, the biggest herd since the early 1930s and far higher than any

149

count since 1949, when wandering wolves, perhaps only a pair, crossed a rare ice bridge from the Ontario mainland and first came to the island. The wolf population at the end of the winter of 1994–95 was 15. The balance between wolves and moose, which was never quite as magical as some popular articles portrayed it to be, was decidedly out of synch.

But Dr. Peterson recently returned from his annual, seven-week-long survey of Isle Royale's two preeminent species, and he reported that the island's wolf population had climbed back up to 22, which was the average in the years before the buildup in the late 1970s to the 1980 peak. Seven pups survived from last summer's litters.

The Isle Royale wolves hunt in separate packs, and, Dr. Peterson said, "There are pups in all three territorial packs for the first time in nine years. The wolves appear to be making a strong comeback." But, he said, they are not out of danger. "The virus disappeared a few years ago, but we haven't resolved the main scientific question," he said. That, he said, is: Does the lack of genetic diversity pose a problem?

The Isle Royale moose population, meantime, has plunged to an estimated 1,200 animals, half the previous year's count. "It was a matter of time until something broke, and it broke in a big way," Dr. Peterson said. "Everything turned against the moose at once. It was the severest winter I've seen on the island. The temperature fell to forty-three degrees below zero, the coldest on record. Three feet of snow aggravated an extreme shortage of winter browse, and the moose also suffered from a heavy winter tick infestation that caused substantial hair loss and left them in a weakened state."

Dr. Peterson described starving moose falling into Lake Superior when they tried to "reach the last tidbit" of browse on trees, leaning from the tops of steep cliffs. "I doubt that any calves, which are weaker than adult moose, will survive the winter," he said. "By February, the proportion of calves in the herd already was down to five percent, lower than in any of the last thirty-eight years." Calves and elderly animals are the main winter prey of the wolves.

"The die-off will make it easier for wolves to bring the moose numbers back under control so the island's overbrowsed forest can begin to recover," Dr. Peterson said. "The wolves couldn't have done it by themselves." He said that a major forest regeneration would occur if the herd were reduced to between 500 and 700 animals.

The sudden crash of the island's moose population was no surprise to experienced wildlife biologists. "It had to happen, given a tough winter and a forage base that was in such poor shape," said Dr. James Peek, professor of wildlife resources at the University of Idaho in Moscow.

"The Isle Royale experience tells us that the balance of nature is a dynamic process, not the status quo," Dr. Peek said. "What we're seeing in the way of predator-prey-vegetation interactions is a constantly changing kaleidoscope of events. The moose herd probably will never increase to that level again. But it will increase if there are mild winters and the wolves get on top and suppress the moose so the forage can improve, or if a big forest fire creates a lot of new habitat," he said.

In his recent book *The Wolves of Isle Royale: A Broken Balance,* published by Willow Creek Press, Dr. Peterson emphasized that "the long record of wolf and moose fluctuations at Isle Royale bears no resemblance to a static balance between predator and prey." From 1959 to the crash of 1980, he wrote, wolf and moose populations appeared to cycle in tandem, with wolves peaking about a decade after moose. "Wolves simply followed trends in their primary prey, moose over ten years old," he said.

Neither moose nor wolves lived on Isle Royale until this century, at least in historic times. Indian tribes hunted, fished, and mined copper on the island for 4,000 years, but archaeologists have found no trace of moose bones at their campsites. Dr. Peterson said that moose, which are strong swimmers but rarely venture out on open ice, first crossed the 20-mile-wide channel from the mainland around 1900.

"On Isle Royale, moose discovered a haven from predators and a virgin food supply," he wrote. "For almost thirty years, nothing held back their increase in numbers. The population bomb exploded in the 1920s and the herd swelled to several thousand." Biologists reported that balsam fir, the main winter food for moose, was in desperate shape and predicted a catastrophe. By 1935, only a few hundred starving moose were left.

The herd eventually recovered because of a fire in 1936 that burned 20 percent of the island and created a new forest, Dr. Peterson said. Then in the winter of 1949, nine years after Isle Royale became a national park, wolves made what was apparently a onetime run from Canada.

"Lake Superior rarely freezes over," said Dr. L. David Mech, a research biologist with the National Biological Service in St. Paul, and a wolf expert.

After Years of Gains, Wolf Population on Isle Royale Plummets

There are always surprises waiting when Dr. Rolf O. Peterson arrives on Isle Royale National Park in Lake Superior to continue a decades-old annual winter study of the wilderness retreat's wolf and moose populations. But the past few years have produced some startling developments, like the recent death of 80 percent of the moose herd. The latest is the crash of a wolf population that had seemed to be on the rebound after a decade of concern about its possible extinction.

In the winter of 1996–97, Dr. Peterson and his colleagues counted 24 wolves on the 210-square-mile island. This was the highest tally since the early 1980s, when an outbreak of deadly canine parvovirus tore through the park's three wolf packs and cut the number of wolves from 50 to a dozen or so. Moreover, 10 wolf pups were spotted last summer, and the biologists expected to find a modest increase in the population when they landed by ski plane in early January for a seven-week stay.

Instead, they discovered that 13 of the older wolves had died and only three pups had survived, leaving a total of just 14.

"We thought the wolves were doing fine," said Dr. Peterson, a wildlife ecologist at Michigan Technological University in Houghton. "The sudden decline could be an aftershock of the moose die-off in the spring of 1996, when we lost nearly two thousand animals, almost eighty percent of the herd. Or it could mean that disease is again present in the wolf population."

Since 1970, Dr. Peterson has continued a study of the relationship between a powerful predator, its primary prey and their environment that was begun in 1958 by the late Dr. Durward L. Allen of Purdue University. It is believed to be the longest-running study of its kind in the world.

When the researchers returned last winter, they found only 500 moose on Isle Royale, the lowest number in 40 years.

"Many of the moose that perished were older animals, the ones that wolves prefer to attack," Dr. Peterson said. "A healthy adult moose can defend itself under most circumstances, so the wolves have had to rely on calves, which were few in number. When food is tight, it's not uncommon for one pack to trespass on another's territory and for the leaders to kill each other.

"What has now happened is that we've moved to a completely new generation of wolves," he continued. "All of them are less than five years old and their reproductive performance will be of great interest since they are even more inbred than their parents."

The Isle Royale moose herd, meanwhile, has increased to some 700 animals that are generally in good shape because of a mild winter and reduced competition for food. The park's battered forest, too, also shows signs of regeneration, Dr. Peterson said.

—LES LINE, April 1998

"It has to be cold enough and calm enough for a bridge to form, and then you need a pack of wolves predisposed to make a long trek over the ice. The chances of that happening are pretty remote or Isle Royale would have had wolves long ago."

In fact, genetic studies using mitochondrial DNA indicate a single ancestor for all of Isle Royale's wolves. Scientists tested almost 250 wolves from the Ontario mainland before finding the genotype in an animal taken by a trapper 50 miles north of Lake Superior, Dr. Peterson reported. "DNA fingerprinting showed the wolves to be highly inbred, comparable to members of a single family. Further analysis revealed that wolves on Isle Royale had lost roughly fifty percent of the genetic variability of mainland wolves." Isle Royale, he said, would provide "an acid test for the notion that inbred populations are not viable."

Scientists point to the cheetah, which has become so inbred from a long-term population decline in Africa that it has less than one-tenth of the genetic variety of domestic cats. Cheetah cubs in zoos and wildlife parks often die before reaching maturity and they are vulnerable to disease. Moreover, studies have found 70 percent of cheetah sperm to be abnormally formed.

While genetic decay may have prevented the wolf population from rebounding, Dr. Peterson believes food was a limiting factor. Few moose from the ages of two to eight are killed by wolves, he noted. "Prime-age moose are too dangerous to approach," he said. "Moose commonly stand and pugnaciously face the wolves, which take the cue and leave. To my knowledge, no one has ever observed wolves killing a moose that did not run when first confronted by its predators." The front legs of a 900-pound moose are formidable weapons. Dr. Peterson described an old and blind bull moose that stood its ground against a pack of wolves for three days until the predators decided to look elsewhere.

Old moose are the wolves' bread and butter, he said, and in recent years the Isle Royale herd has been dominated by vigorous young adult moose. "In the late 1990s," he wrote, "alpha positions in wolf packs will be filled by a new generation of young wolves, disease-free and well-fed throughout their lives. Their reproductive success should shed light on the important issue of genetic variability in small, isolated populations."

Dr. Mech, who was Dr. Allen's first wolf student almost 40 years ago, said Isle Royale "represents the best experiment to determine the effects of inbreeding on a wild population of canids, and it is showing wolves to be much more resistant to inbreeding depression than one might expect." But Dr. Robert Lacy, a population geneticist at the Brookfield Zoo in Chicago, believes that view is "too optimistic."

"There's a lot of chance involved in how inbreeding affects any population," Dr. Lacy said. "A number of studies have shown that one small group of animals will get lucky and do well and another group will get into a lot of trouble with developmental defects and high infant mortality. Maybe the Isle Royale wolves just got lucky."

A classic case, Dr. Lacy noted, is the small surviving population of endangered Florida panthers. "The effects of inbreeding on the panthers are severe. Most of the males born in recent years have either one testicle or none, their sperm is poor and many of the kits show congenital heart defects," he said. Biologists have introduced cougars from east Texas in the Everglades to enhance the panther gene pool.

"Another reason I'm pessimistic is that the effects of inbreeding may not show up until the population is stressed by food shortages or a new disease hits," Dr. Lacy said. "The news is hopeful, but I'm still worried about the future of the Isle Royale wolves."

—LES LINE, March 1996

5

SEAFARING MAMMALS

Mammals seem to have a deep affinity for water, judging by the number of species that have independently found ways to make an aquatic living. Hippos, otters and beavers spend most of their time in the water. And three orders of mammals have adapted their bodies for living completely or almost completely in the sea. Seals use the land only for breeding. The sirenians (dugongs and manatees) live completely in water, inhabiting coastlines and estuaries. And the cetaceans—whales and dolphins—have mastered life in the far oceans.

The three groups of marine mammals—sirenians, whales and seals—differ widely in character. The sirenians are thought to be distant relatives of the elephant, but as far as is known they lack the pachyderm's prodigious memory.

New finds of fossil whales confirm the suspicion that cetaceans evolved from animals similar to today's hooved animals, probably an ungulate group called the mesonychids. Whales seem to be more intelligent than cows and horses, and their cousins the dolphins appear to be among the brightest mammals.

Adapting to life in the sea has entailed major changes in the mammalian body plan. All of the three kinds of marine mammals shave streamlined bodies, but otherwise each is different because of the different species from which they are derived. Whales have turned their forelimbs into flippers and have lost their hind limbs and all but the hint of a pelvis.

The fossil history of marine mammals is particularly incomplete because they do not fossilize well. Corpses in the sea are quickly dismembered and few sink intact into sediment, where the slow process of fossilization can take place. Paleontologists are uncertain as to whether the kinds of whale (toothed whales and baleen whales) arose independently from terrestrial ancestors or from some early split in the cetacean heritage.

The ancestry of seals is also somewhat vague. There are three different families: the eared seals, the walruses and the earless seals. Both walruses and the eared seals, which include sea lions and the fur seal, are thought to have evolved from a bear-like ancestor. The earless seals are so different that they seem to have evolved independently from some other kind of Carnivora, perhaps a member of the weasel family.

Dolphin Courtship: Brutal, Cunning and Complex

AS MUCH AS PUPPIES or pandas or even children, dolphins are universally beloved. They seem to cavort and frolic at the least provocation, their mouths are fixed in what looks like a state of perpetual merriment, and their behavior and enormous brains suggest an intelligence approaching that of humans—or even, some might argue, surpassing it.

Dolphins are turning out to be exceedingly clever, but not in the loving, utopian socialist manner that sentimental Flipperphiles might have hoped. Researchers who have spent thousands of hours observing the behavior of bottlenose dolphins off the coast of Australia have discovered that the males form social alliances with one another that are far more sophisticated and devious than any seen in animals apart from human beings. They have found that one team of dolphins will recruit the help of another team of males to gang up against a third group, a sort of multitiered battle plan that scientists said requires considerable mental calculus to work out.

But the purpose of these complex alliances is not exactly sportive. Males collude with their peers as a way of stealing fertile females from competing dolphin bands. And after they have succeeded in spiriting a female away, the males remain in their tight-knit group to ensure the female stays in line, performing a series of feats that are at once spectacular and threatening. Two or three males will surround the female, leaping and belly flopping, swiveling and somersaulting, all in perfect synchrony with one another. Should the female be so unimpressed by the choreography as to attempt to flee, the males will chase after her, bite her, slap her with their fins or slam into her with their bodies. The scientists call this effort to control females "herding," but they acknowledge that the word does not convey the aggressiveness of the act.

"Sometimes the female is obviously trying to escape, and the noises start to sound like they're hurting each other," said Dr. Rachel A. Smolker of the University of Michigan in Ann Arbor. "The hitting sounds really hard, and the female may end up with tooth-rake marks."

Dr. Smolker, Dr. Andrew F. Richards and Dr. Richard C. Connor, who is now at the Woods Hole Oceanographic Institution in Massachusetts, reported their findings about dolphin alliances and herding in *The Proceedings of the National Academy of Sciences*.

The researchers said that while marine biologists have long been impressed with the intelligence and social complexity of bottlenose dolphins—the type of porpoise often used in marine mammal shows because they are so responsive to trainers—they were nonetheless surprised by the intricacy of the males' machinations. Many male primates, including chimpanzees and baboons, are known to form into gangs to attack rival camps, but scientists have never before seen one group of animals soliciting a second to go after a third. More impressive, the two-part alliances among dolphins seem to be extremely flexible, shifting from day to day depending on the dolphins' needs, whether or not one group owes a favor to another, and the dolphins' perceptions of what they can get away with.

The creatures seem to be highly opportunistic, which means that each animal must always be computing who is friend and who is foe. "If you think of an interaction between groups that is predictably hostile, it doesn't seem to require much gray matter to know where you stand," said Dr. Connor. "But when you have situations always changing between alliances, you get the soap opera effect. 'What did he do with her today?' 'Should we go after them tomorrow?' "

The biologists also have evidence that females form sophisticated alliances in an effort to thwart male encroachment, and that bands of females will chase after an alliance of males that has stolen one of their friends from the fold. What is more, females seem to exert choice over the males that seek to herd them, sometimes swimming alongside them in apparent contentment, but at other times working furiously to escape, and often succeeding. But female dolphin behavior is usually more subtle than the male theatrics, and hence less easily deciphered, particularly under the difficult field conditions of studying animals that spend much of their time underwater.

Dr. Connor and others suggest that the demands of intricate and changing social allegiances and counterallegiances could have been the force driving the evolution of intelligence among dolphins.

"The smarter some animals get, and the greater their ability to form and use alliances, the more important it is for other animals to get as smart," said Dr. Richard W. Wrangham, a professor of anthropology at Harvard University who has studied social behavior among primates. "This could be the sort of selective pressure one is looking for to explain the evolution of the dolphin's brain."

Lest it seem that dolphins are little more than thugs with fins and a blowhole, biologists emphasize that they are in general remarkably good-natured animals, and usually live up to their reputation as sportive, easy-going and communal.

"When you put them into a captive situation, they're like little puppy dogs," said Dr. Kenneth Norris, professor emeritus of the University of California at Santa Cruz and one of the world's authorities on dolphin behavior. "Sure, sometimes they'll bite, but it's not like trying to train a leopard. They're orders of magnitude more peaceful than that."

Most of the 30 species of dolphins and small whales are extremely social, forming into schools of several to hundreds of mammals, which periodically break off into smaller clans and then come back together again in what is called a fission-fusion society. Among other things, their sociality seems to help them evade sharks and to forage for fish more effectively.

Species like the bottlenose and the spinner dolphins make most of their decisions by consensus, spending hours dawdling in a protected bay, nuzzling one another and generating an eerie nautical symphony of squeaks, whistles, barks, twangs and clicks. The noises crescendo ever louder, until they reach a pitch that apparently indicates the vote is unanimous and it is time to take action—say, to go out and fish. "When they're coordinating their decisions, it's like an orchestra tuning up, and it gets more impassioned and more rhythmic," said Dr. Norris. "Democracy takes time, and they spend hours every day making decisions."

As extraordinary as the music is, scientists have not found evidence that dolphins possess what can rightfully be called a complex language, where a dolphin can clearly say to another, "Let's go fishing." "We've yet to come up with much context that is specific to any of the sounds," said Dr. Norris.

But the vocalizations are not completely random. Researchers have determined that each bottlenose dolphin appears to have its own call sign—a signature whistle unique to that creature. Whistles are generated internally and sound more like a radio signal than a human whistle. The mother seems to teach her calf what its whistle will be by repeating the sound over and over. The calf retains that whistle, squealing it out at times as though declaring its presence. More impressive, one dolphin may occasionally imitate the whistle of a companion, in essence calling the friend's name.

But dolphin researchers warn against glorifying dolphins beyond the realms of mammaldom. "Everybody who's done research in the field is tired of dolphin lovers who believe these creatures are floating hobbits," said Karen Pryor, a dolphin trainer and scientist who lives in North Bend, Washington. "A dolphin is a healthy social mammal, and it behaves like one, including doing things that we don't find particularly charming." Ms. Pryor and Dr. Norris have edited a book that sums up the state of the dolphin field, called *Dolphin Societies: Discoveries and Puzzles,* published by the University of California Press.

Dolphins become particularly churlish when they want to mate, or to avoid being mated. Female bottlenose dolphins bear a single calf only once every four or five years, so a fertile female is a prized commodity to the males. Because there is almost no size difference between the sexes, a single female cannot be forced to mate by a lone male. That may be part of the reason why males team into gangs.

In the latest research on bottlenose dolphins, Dr. Connor and his colleagues spent the last 10 years studying a network of about 300 dolphins in Shark Bay, in western Australia, and devoted 25 months to observing male behavior in detail. They followed dolphins around in a 12-foot dinghy, identifying individuals through scar patterns on their fins and recording their whistles and clicks whenever possible.

The researchers have discovered that early in adolescence, a male bottlenose will form an unshakable alliance with one or two other males. These dolphins stick together for years and perhaps a lifetime, swimming, fishing and playing together, and flaunting their fast friendship by always traveling abreast and surfacing in exact synchrony.

Sometimes that simple pair or triplet is able to woo a fertile female on its own, although what happens once the males have herded in a female, and whether she goes for one or all of them, is not yet known: The researchers have yet to witness a dolphin copulation. Nor do researchers understand how the males determine that a female is in estrus or at least approaching it, and thus is worth attempting to herd. Males do sometimes sniff around a female's genitals, as though trying to sense her receptivity, but because bottlenose dolphins give birth so rarely, males may attempt to keep a female around even when she is not ovulating, with the hope that she will require their services when the prized moment of estrus arrives.

At other times potential mates are scarce, and male alliances grow obstreperous. That is when pairs or triplets may seek to steal females from other groups. To do that, they seek out another alliance of lonely bachelors, and somehow persuade that pair or triplet of dolphins to join in the venture. The researchers are not yet sure what signals the males use to recruit outside aid, but they believe the supplicants use their pectoral fins to stroke the males from which they need assistance, or perhaps give them a few gentle pecks.

In simpler maneuvers among primates, scientists have observed that when one male needs the help of another, he takes a rather blunt approach.

"In baboons, a male who wants help against an enemy will look first at his friend, and then let his eyes trail over to the enemy, flicking his eyes back and forth," said Dr. Wrangham.

However the pact is sealed, the two dolphin gangs will then descend on a third group that is herding along a female. The two groups will then chase and assault the defending team, and because there are more of them, they usually win, taking away the female. Significantly, the victorious joint alliance then splits up, with only one pair or triplet getting the female and the other team apparently having helped them strictly as a favor.

That buddy-buddy spirit, however, may be fleeting. Two groups of dolphins that cooperated one week may be adversaries the next, as a pair of males switches sides to help a second group of dolphins pilfer the same female they had helped the now defending males capture in the first place.

How many of these encounters involve relatives ganging up against nonrelatives is not yet known. The researchers hope soon to begin doing DNA fingerprinting on the dolphins to determine family trees.

In the strongest evidence of two-tiered coalitions, the researchers at one point watched as a pair of males approached another alliance that was herding a female, but did nothing at the time.

"They watched for a while, and then they swam away," said Dr. Connor. "But later they came back with another alliance to attack them and capture the female."

The instability and intricacy of the mating games may explain why males are so aggressive and demanding toward the females they do manage to capture. Male pairs or triplets guard the female ferociously, jerking their heads at her, charging her, biting her, and leaping and swimming about her in perfect unison, as though turning their bodies into fences. They may swim up under her, their penises extruded and erect but without attempts at penetration. Sometimes a male will make a distinctive popping noise at the female, a vocalization that sounds like a fist rapping on hollow wood. The noise seems to indicate "Get over here!" because if the female ignores the pop, the male will threaten or attack her.

At some point, the female mates with one or more of the males, and once she gives birth the alliance loses interest in her. Female dolphins raise their calves as single mothers for four to five years.

Having mapped out the basics of male alliances, the researchers are now trying to better understand female social behavior. "Our research has been male-centered because it's easy," said Dr. Smolker. "Males make big movements and it's clear what's going on. But females must be playing a critical role."

The scientists said females seem to have widely varying habits. Most males form into lifelong pairs or triplets, but females may or may not ally themselves with friends.

"Some females are solitary and forage alone, some have stable relationships with a few other females, and some are all over the place, like social butterflies," said Dr. Connor.

Biologists suggest that the pressure to alternately cooperate and compete with their fellows may have spurred on the evolution of the dolphin brain. Dolphins have one of the highest ratios of brain size to body mass in the animal kingdom; this is often a measure of intelligence.

A similar hypothesis has been proposed for the flowering of intelligence in humans, another big-brained species. Like dolphins, humans evolved in highly social conditions where kin, friends and foes are all mingled together, and the resources an individual could afford to share today might become dangerously scarce tomorrow, igniting conflict. In such a setting, few relationships are black or white, and the capacity to distinguish subtle shades of gray demands intelligence.

But scientists concede that a big brain might have evolved first, and the sophisticated social behavior developed later. They point out that sheep, which are hardly known for their savoir faire, also possess unusually large brains.

—NATALIE ANGIER, February 1992

Sea Lions to Report on
Behavior of Whales

WHEN JENIFER ZELIGS strolls past the pools where the California sea lions are kept, the animals haul themselves out of the water and waddle toward her, their snouts snuffling in excitement, their blubbery bellies quivering like buckets of chocolate pudding.

They bark to get her attention, but they are not barking for fish; the sea lions have already been fed. No, they want her, their blond trainer, their camp counselor, their bipedal and terrestrial best friend. They love Ms. Zeligs because she plays games with them. She shows them new tricks. She scratches their fur, grooms them, cleans their teeth and takes them for walks around their enclosures.

"I make them feel good about themselves," she said. "They get training sessions every day. If you didn't have a training session, you'd be depriving the animal of what makes it most happy in life."

In return, the four sea lions that live at the Long Marine Lab at the University of California at Santa Cruz will do almost anything Ms. Zeligs asks of them. Soon she will be asking quite a bit, perhaps more than a scientist has ever requested of a pinniped. She will ask the sea lions to help perform basic research, both on themselves and on some of the other creatures they encounter in the wild.

Ms. Zeligs, along with Dr. Daniel P. Costa, her thesis advisor and an associate professor of biology at the university, and Dr. James T. Harvey of the Moss Landing Marine Mammal Center in California, have devised a novel and potentially revolutionary scheme for studying the physiology and behavior of marine mammals. They are training sea lions to do lengthy dives in the open ocean on command, and to take movies while underwa-

165

Patricia J. Wynne

ter of whales, dolphins and other species that are exceptionally difficult to study by conventional means.

The scientists are teaching the sea lions to carry sophisticated equipment on their backs and to obey diving and swimming instructions so precisely that the sea lions should be able to aim video cameras at whales and gather footage of the secretive giants at play, work, love or battle.

The project is at its earliest stages, but the sea lions are swiftly learning to follow many intricate hand and audio signals. They may be ready to begin doing test filming in their tanks by the end of the summer and to start filming gray whales along the Santa Cruz coast by the middle of next year.

The use of sea lions as spies in the ocean wilderness has several advantages over either human divers or submersible vehicles like the famed research vessel *Alvin*. Whales often react negatively to foreign objects like submarines or people in wet suits, but they are quite accustomed to sea lions, which frequently skate along in their wake to save energy or swim around them in the search for food.

Sea lions can also descend comfortably to depths that are forbidding and expensive for people to plumb; and while *Alvin* is notoriously slow and clumsy to maneuver underwater, a sea lion can smoothly zig and zag to trail an active whale wherever it may dip or tumble. Sea lions are easily trained, which is why they are staples of ocean aquariums, where they are often seen balancing balls on the ends of their noses.

"The idea behind this is that usually, when you study whales by introducing artificial things into their environment, you change their behavior," said Dr. Harvey. "But whales are used to seeing sea lions, so by getting sea lions to be our filmmakers, we should be able to record what the whales are doing without inadvertently influencing what they are doing.

"We've got a biased view of what whales are doing because most observations and reports have been done from shipboard. In reality, whales spend only five to ten percent of their time at the surface. We need to know what's going on underwater."

In another phase of the project, the biologists will measure the performance of the sea lions themselves to learn how the burly creatures manage to plunge hundreds of feet below sea level and then thrash about in a vigorous search for food, all the while holding their breath and expending so many calories that scientists wonder why the animal does not simply perish on the spot from overwork.

To solve the puzzle, the researchers will ask the animals to simulate their natural diving and foraging patterns in the ocean, but to do so in such a focused and well-timed fashion that the scientists can track precisely what is going on in the animal's body from one dive to the next.

The biologists will measure revealing details like the creatures' oxygen consumption and the flux in their body temperature and heart rate before and after their deepest dives. They will also be looking for any changes in blood gases that could explain why the rapidly resurfacing sea lions do not suffer from the bends—a painful condition, caused by nitrogen bubbling up in the blood when human divers ascend too quickly, that can result in death.

By analyzing the details of sea lion gymnastics in the open ocean, the scientists hope to gain insights into the remarkable adaptations displayed by all marine mammals, which are poorly understood. That basic knowledge will in turn allow them to understand how the mammals fit into the ocean habitat and how their lives are disturbed by natural events, like an El Niño current, as well by as unnatural stresses, like human fishing practices.

Two of the four sea lions in Dr. Costa's project already have experience working for humans. Those animals were donated to the Santa Cruz laboratory by the Naval Ocean Systems Center in Hawaii, where the sea lions were used to help recover military instruments and weapons from the ocean floor.

Ms. Zeligs and a team of volunteers are learning that sea lions will heed a broad range of commands if given enough motivation and encouragement, and if they are kept sufficiently entertained. Thus far, the researchers have trained the animals to wear a variety of monitors and devices, dive to specific depths and stay underwater until told to resurface. Before and after each dive, the animal breathes into a metabolic chamber floating at the surface of the water, and the chamber analyzes the amount of oxygen used in the dive.

Right now, the training and initial physiological studies are taking place at the Long Marine Laboratory, where the sea lions swim in tanks that are 15 feet deep. But the researchers hope to move the animals to pens beside the ocean by this fall. From there it will be a simple chore to take them out into the ocean depths and ask that they make far deeper and more complex dives.

The scientists also plan to begin constructing a life-size model of a California gray whale, a plastic hulk that will be placed over a boat or a small submarine and used to train the sea lions to follow specific features of the whale. The sea lions may be asked to swim beside the mock whale or under the whale, or to swim beside the model's eye or mouth. By positioning the camera on the sea lion's back in a specific orientation, the scientists hope to be able to get footage of whale calves nursing, for example, or whales mating, events that have been notoriously difficult to observe.

The animals expected to meet such grand expectations are among the most common of marine mammals. In San Francisco, sea lions loiter around the public piers, where tourists gleefully toss them fish. Elsewhere, sea lions have learned to steal fish from hatcheries and fish runs, and, as a result, they are sometimes shot or bludgeoned to death by fishermen, although the practice is illegal.

Shielded by the Marine Mammal Protection Act of 1972, the California sea lion, like some other marine mammals, has thrived, and now there are at least 150,000 along the Pacific coast of North America. They are rambunctious, hulking creatures, with the adult females weighing about 250 pounds and the males almost three times that weight.

Unlike most seals, sea lions have external ears and a long rear flipper that allows them to move around more easily on land. They eat 20 to 30 pounds of fish a day, repeatedly diving down to about 700 feet and staying there for up to nine minutes a time.

Dr. Costa and Ms. Zeligs initially conceived of their project as a way of studying sea lion adaptations in a far more sophisticated manner than anybody had managed before. They decided to train sea lions to recapitulate their normal diving and foraging behavior on command so the researchers could pick apart revealing physical changes, element by element, as the animals performed in the ocean.

If they could learn how much oxygen the sea lion consumed while diving, for example, they could calculate through standard metabolic equations precisely how much energy was needed to get the animal down to a given depth for a given period of time. Similarly, measurements of the animal's core body temperature when it reached the deepest point of its dive would reveal whether a sea lion adapted to the rigors of the ocean by turning down its thermostat, a potential way of saving energy.

Then Dr. Harvey suggested that, beyond using sea lions to study mammalian physiology, the trained animals could be used to explore other species as well.

Ms. Zeligs has been training sea lions, monkeys and other animals since the age of 12 at the National Zoo and elsewhere. She is in charge of teaching sea lions to do her bidding. Many people have compared the intelligence of the animals to that of dogs, but she said sea lions were brighter by far.

"They have a much better understanding of the complex relationships among variables," she said. Ms. Zeligs, like other accomplished animal trainers, has learned that food is a poor motivator to train a clever creature like a sea lion. Instead, Ms. Zeligs has made herself and her corps of volunteers into the most stimulating part of the sea lions' lives.

The trainers spend at least 20 minutes three times a day with each sea lion, playing games and teaching them new skills, all with the object of motivating the sea lions to enjoy listening to humans. The amusements vary from animal to animal and from day to day.

"Some love water games; some go crazy for Frisbees or Wiffle balls," said Ms. Zeligs. "Wiffle balls have turned out to be one of our most incredible motivating factors. If you offer the sea lions fish or a Wiffle ball, they'll go for the Wiffle ball every time."

The animals also like being petted and scratched, particularly during molting season, and they willingly submit themselves to regular checkups and the taking of blood samples.

"It's the best possible health plan a sea lion can imagine," Ms. Zeligs said. The trainers are careful to vary the routine to keep the sea lions from getting bored or slothful.

As a result of the loving, if laborious, care, the sea lions will obey any number of commands. They will follow a little anchor down to a given depth, touch their snout to the anchor and keep it there until a little bell goes off, telling them to surface. They will swim in a particular pattern alongside a trainer or one another, a necessary skill if they are to manage to follow a roaming whale eventually.

The researchers have begun putting harnesses on the sea lions, along with backpacks that will carry heart-rate monitors, thermometers and, eventually, the housing for a 28-ounce video camera. To keep the camera steady, it will be placed on gimbals, the same device that keeps a ship's compass in a horizontal position.

Assuming that the rehearsals with the whale model succeed, the trainers will get the sea lions working on the real thing. Using hand signals, they will ask the sea lions to find a whale, then swim beside it for 30 minutes from a distance of about 30 feet. The first films will be taken of gray whales, which abound along the Pacific coast, but the scientists see no reason why the sea lions could not learn to follow humpback whales, right whales or even the blue whale, the mightiest creature ever to grace the Earth, and perhaps the least understood.

—Natalie Angier, July 1992

Hearing of Manatees May Prove to Be Key to Protecting Species

AT THE FLASH OF A HAND signal in the water, Stormy, a 1,200-pound, seven-year-old manatee, swam to place his head in the wire hoop suspended in his tank. Seconds later, a light came on at the end of the tank, signaling that it was time to swim out of the hoop and bump his lips against a paddle suspended to the right of the light. When he responded correctly, his trainer blew a whistle that told him to collect his reward, a monkey biscuit.

The behavior Stormy was learning will soon be used to develop a manatee hearing test. Stormy will swim to insert his head in the hoop, and a tone will be sounded in the water. When the light comes on, Stormy will indicate if he heard the tone by swimming to the left paddle if he did and to the right if he did not.

Six months ago, Dr. Edmund Gerstein of Florida Atlantic University began Stormy's training here at Lowry Park Zoo in Tampa. When Dr. Gerstein advances to the hearing test, he will be studying Stormy and another manatee, Dundee, to try to find out why nearly every manatee spotted in Florida's waters seems to have collided with a boat. Could it be that manatees cannot hear approaching boats so they are unable to swim out of harm's way?

The research is part of an effort to protect manatees, seal-like marine mammals sometimes called sea cows, that grow up to 13 feet long and weigh up to 3,500 pounds. Although they have been on the endangered species list for nearly two decades, protection programs have been slow to develop. Under a recovery plan for manatees that got under way in 1983, federal and state officials are doing basic research to determine their life history and habitat requirements, and are taking steps to protect their habitats, reduce mortality and educate the public.

But no one can say for sure whether the plan will work. "We're playing catch-up right now," said Patrick Rose, the administrator of the Office of Protected Species in the Florida Department of Natural Resources.

Scientists say that at least 1,800 West Indian manatees roam the coasts and coastal rivers of Florida and southern Georgia, but, given the difficulty of spotting the animals under murky water, no one is certain how many there are, or whether the population is increasing or decreasing.

In 1991, at least 174 manatees died in Florida. Fifty-three of those died of injuries caused by boats, according to the Department of Natural Resources, which determines the cause of death in each case. The boat-related deaths set a new record, but just barely. The death toll has been climbing for several years, and scientists are not sure about how long the population can withstand such a high mortality rate.

"Right now it's anybody's guess whether the number dying exceeds the number being born," Mr. Rose said. "Regardless of where the population is today, the animal's ability to recover gets more difficult every day.

"Some see the manatee as a warm, cuddly issue," he said. But, he added, "They are the barometer of how well we are able to protect coastal ecosystems. As the manatees fare, so do systems."

The research on the manatees' hearing was originated by Geoffrey Patton, a senior biologist at Mote Marine Laboratory, a nonprofit research organization in Sarasota. He is collaborating with Dr. Gerstein.

"One manatee in Florida has been hit at least twelve times, judging from his scars," Mr. Patton said. "Why don't they learn to avoid boats?"

Mr. Patton and Dr. Gerstein say the solution may be to learn to warn the animals of oncoming danger. Once they determine the manatees' hearing range, they will try to find out how well manatees hear in the presence of background noise and whether they can tell the direction of the tone.

"If we could determine their hearing ability, it may be possible to modify the sounds that boats make so the animals can locate them and get out of the way," Mr. Patton said. "It could boil down to some simple plastic device on the hull that would vibrate at the right frequency and cue the animal."

A few feet away from Stormy's tank, two wild manatees were recovering from their injuries under the watchful eye of zoo volunteers. One had a collapsed lung that had probably been pierced by a rib broken in a boat collision. The other had been caught in a crab-trap line, which had cut off the circulation in his flipper. Both are expected to recover.

The manatees of Florida, which are genetically distinct from a small population of manatees found in the Caribbean and South America, are protected under a state law that dates back to 1893, as well as by the 1972 Marine Mammal Protection Act and the Endangered Species Act of 1973.

During the 1970s, however, relatively little was done to help the manatees, primarily because of a lack of money, staff and commitment from both the Fish and Wildlife Service and the state Department of Natural Resources. Protection efforts increased in the 1980s but were still insufficient, so, at the urging of the Marine Mammal Commission, a federal advisory group, a new recovery plan was written in 1989. In that same year, Florida established the Save the Manatee Trust Fund, which relies on the public to make donations and purchase special manatee license plates to finance the state's program.

One priority of the recovery plan is to determine the manatees' life history and habitat requirements. Since the late 1970s, scientists have conducted aerial surveys to estimate populations at specific locations, for example, near the warm-water outfalls of power plants and the natural hot springs, where the manatees congregate in winter. But those surveys opened "only a very narrow window into their lives," said Dr. Tom O'Shea, director of the Sirenia Project, the manatee research agency of the Fish and Wildlife Service.

The federal agency and the state are now using tracking techniques that allow them to follow an animal continuously for weeks and months. About 20 wild manatees have been fitted with harnesses above their tails; a stiff tether extending from the harness holds an electronic transmitter above the waterline. The electronic signals are picked up by satellite, enabling scientists to find out where these animals congregate and how long they stay there. That information, along with the findings from manatee carcasses and other data, is used to make decisions on which areas are essential to the manatees.

The service is also keeping track of 900 manatees that can be identified by the distinctive scar patterns left from their collisions with boats. Each winter, when the manatees congregate at warm-water sources, biologists identify individuals, take photographs and try to determine the condition and reproductive status of each animal. The information is used to gauge birth and death rates and migration patterns.

The injured manatees at the Lowry Park Zoo were being cared for under the recovery plan's mandate to rescue and rehabilitate as many injured or diseased manatees as possible. Rescue teams transport injured animals to one of five ocean areas, like the Miami Seaquarium, where marine mammal veterinarians try to nurse them back to health.

"Our goal is to release as many of those animals back into the wild as we can so they can reproduce naturally," said Robert Turner, the manatee-recovery coordinator for the Fish and Wildlife Service. In May, three manatees were released at the Merritt Island National Wildlife Refuge, including a female who had been injured and her calf, which was born in captivity. Most released manatees are fitted with transmitter harnesses so their progress can be tracked.

Although the survival rate for released manatees is good, Mr. Turner said, orphaned calves and rehabilitated manatees who have spent long periods in captivity may not know where to go or how to get food. "It's like taking a pet and throwing it into the woods," he said. To improve survival rates for such animals, the service plans to build a special pen at the refuge, sort of a halfway house, where manatees that are ready for release can become acquainted with wild manatees across the fence until they are ready to be turned loose.

The most controversial aspect of the manatee program is the attempt to slow boats down near manatee habitats. The Department of Natural Resources has set speed limits for hundreds of miles of waterways in nine counties and plans new speed limits for four more. The regulations, which are imposed after consultations with the counties, require boats to go very slowly in some manatee gathering spots and travel corridors, and they prohibit all human activities in some areas.

While manatee advocates endorse the plan, many water skiers and boaters have fought the rules. In Sarasota County, speed zones were adopted last December, and signs were posted in the waterways in July. But even before the signs were posted, the County Commission had decided to review the regulations and consider suggesting changes to the state.

Rick Rawlins, owner of the Highland Park Fish Camp in Volusia County, said the rules adopted there in 1991 would put him out of business because the slow-speed zones lengthened the time it took to reach fishing spots.

"The rules would add as much as five to six hours to a day of fishing," Mr. Rawlins said. "My customers are leaving me. Some said they are not going to fish anymore. Others are going to other counties."

Mr. Rawlins has formed a group called Citizens for Responsible Boating to fight the regulations. The group filed an administrative appeal with the state; when that was turned down, it filed suit against the Department of Natural Resources, arguing that the economic impact on local businesses had not been fully considered. That suit is pending.

The boating speed limits are not the final step. Each county is also required to develop a comprehensive plan for manatee protection. Each plan must address issues like controls on marina sites and other development, and the plans could include more stringent speed regulations.

Despite the increased efforts of recent years, no one is certain that the manatee will thrive in future years. "We'll be able to tell something once the manatee-protection plan starts taking effect," Mr. Turner said. "If we start to see mortality decline, then I have good hope that we can do something. If, after all these efforts, we still see increases in mortality, I really don't know what the next step will be."

—Catherine Dold, August 1992

Picky Eaters in Monterey Bay
Who Dabble in Petty Theft

SUCKERFACE the sea otter prefers octopus and purple sea urchins for dinner. Another otter, Flathead, likes to chew on crabs and suck the gonads of sea stars, while Nosebuster munches exclusively on turban snails.

Marine biologists at the Monterey Bay Aquarium in California have discovered that sea otters are picky eaters, generally preferring only two or three kinds of prey out of 30, and the preference is largely influenced by their mothers. They also found that male sea otters are prodigious thieves that pilfer a third of their food from the females; the females seem to tolerate this peccadillo.

Led by Dr. Marianne Riedman, director of the Sea Otter Field Research Program at the aquarium, the team of marine biologists tracked more than 60 sea otters along the Monterey Bay Peninsula from 1985 to 1992 to study their role in the coastal marine ecosystem. The studies included observation of the otters' social and mating behaviors, diets, survival strategies and patterns of communication.

"This is the first long-term behavioral observation of sea otters," Dr. Riedman said. "We found that they have great environmental and economical impacts and help maintain the near-shore marine ecosystem."

Whiskered and with a weasel-like face, at up to four feet in length they are the largest mustelids, a group that includes freshwater otters, weasels, minks, skunks and badgers. Scientists believe the sea otters evolved from a primitive fish-eating otterlike animal 5 million to 7 million years ago.

Although newcomers to the sea compared with other marine mammals, they are among the most intelligent and innovative. Dr. Riedman said they are quick to learn from one another and have mastered the use of rocks to crack open shellfish.

Lacking blubber like other marine mammals such as whales and dolphins, sea otters eat up to 30 percent of their body weight daily to keep warm in the chilly waters. But their voracious appetites do not create a food shortage because each otter's preference for a few types of foods means that no single food source is likely to be destroyed in a colony.

When not busy diving for food, the otters wrap strands of kelp around themselves and float on their backs in groups, their paws pointed skyward as if praying. Lifting their paws out of the water helps them conserve body heat, Dr. Riedman said.

The sea otters also keep warm with unusually thick and luxurious double-layered fur, the densest among animals. The soft and fuzzy outer hair forms a protective covering that keeps the fine and dense underfur dry. One square inch of sea otter underfur contains up to one million hairs.

But this feature essential to their survival almost led to their demise as humans discovered their fur and coveted it for their own warmth.

Sea otters once thrived across the rim of the North Pacific from northern Japan to the Alaska Peninsula and along the Pacific coast of North America to Baja California. They are especially plentiful off the Channel Islands of Southern California and the central and northern areas of the state. There were once thought to be nearly 20,000 California sea otters and 150,000 to 300,000 worldwide.

But commercial hunting that began 300 years ago drastically reduced their numbers. By the early 1900s, the sea otters had dwindled to 2,000 worldwide. They are now classified as "depleted" under the Marine Mammal Protection Act of 1972, "threatened" under the federal Endangered Species Act of 1973 and are fully protected in California.

Under government protection, the sea otter population has resurged to 150,000 worldwide. There are now about 2,200 otters along the central California coast, Dr. Riedman said. The current clan of California otters descended from a group of 32 found in Point Sur in 1915. Biologists at first kept their discovery a secret to prevent poachers from hunting them.

While elated by the otters' return, scientists are concerned about the California otters' population growth of 5 percent a year, which lags far behind the rate of 17 to 20 percent among Alaska otters. Dr. Riedman attributes this to the mortality rate of nearly 40 percent for otter pups in California. Drownings in gill nets and other fishing nets along the coastlines also caused California sea otters to decline in the 1970s.

Sea otters are also sensitive to pollution. An oil or gas spill could endanger much of their population, therefore disturbing the delicate ecological balance in the ocean, Dr. Riedman said.

Sea otters help maintain this balance by protecting kelp forests, which provide shelter and nutrients for plants and animals, by feasting on invertebrates like sea urchins and abalones that destroy the kelp, Dr. Riedman said. Kelp also benefits humans. Extracts made from it can be used to make hundreds of products, ranging from ice cream to paint to dissolvable surgical thread.

—JANE J. LII, March 1994

Paternity Tests Unsully Wild Reputation of Faithful Gray Seal

BY ALL PREDICTIONS, the gray seal, found throughout the North Atlantic, should be a classically polygynous mammal, a species founded on the principle of a harem. Males haul themselves up onto breeding beaches where females abound and fight viciously to form a dominance hierarchy, the biggest, nastiest sandkickers monopolizing the greatest number of females. Though under constant threat of overthrow by lesser bulls, the alpha males presumably get their payoff in siring the most pups.

For their part, the females mate with whatever big boy is defending the turf closest to her; and given the fragility of seal power, her partner changes from season to season.

This, at least, was the theory, one buttressed by abundant behavioral observations. Alpha males were often spotted copulating with the females in the territory they commanded. However, when scientists from the University of Cambridge decided to do some paternity testing of gray seals using molecular genetic techniques, the results shook them up on two fronts.

To begin with, they determined that the most dominant males at their study site on North Rona, a small island off northwest Scotland, fathered shockingly few of the pups born on the island each season.

More surprising still, the scientists discovered that a large number of pups born in consecutive years to the same females were in fact full siblings—not half siblings as one would expect if the mother mated with a different male each season. In sum, the scientists found evidence of mate fidelity in a species one might otherwise consign to mammalian Gomorrah. The females appeared to be mating with the same male from year to year, but where those males are, and when the females couple with them, remains unknown. Very few of the paternities could be assigned to any of

180

remains unknown. Very few of the paternities could be assigned to any of the males flopping around North Rona island.

"The results were quite unexpected," said Dr. Bill Amos. "This means we have to cast a skeptical eye on a lot of our assumptions and test other polygynous mammals" to check who is fathering whom.

Dr. Amos reported the results in the journal *Science* with his colleagues Sean Twiss, Paddy Pomeroy and Sheila Anderson.

The mammals are by no means monogamous. In only 30 percent of the cases they tested were pups born in consecutive years full siblings. Yet that percentage is something close to 30 percent higher than what the scientists might have expected. And if they did find full siblinghood among the pups, the researchers would have predicted the father would be a familiar big male on beach, because males often retain dominance through several mating seasons.

Of the 48 pairs of pups they tested genetically, however, only two pairs were the offspring of resident dominant males—a pathetic success rate, said Dr. Amos, that one might expect by chance alone. Clearly the female seals were not abiding by the terms of harem life, and instead were doing their dallying elsewhere, presumably—at least about a third of the time—with a rough proximation of a spouse.

These results are as remarkable in their way as those of several years ago, when molecular techniques revealed that supposedly monogamous birds were anything but. In this case, biologists found an unforeseen reservoir of fidelity among mammals presumed to be fickle.

"We must conclude that many pairs of seals establish durable ties, recognizing each other between seasons and coordinating their behaviors," the authors write.

Other researchers who study animal mating behavior were at a loss to understand the new results. Dr. Tim Clutton-Brock of the University of Cambridge, who did pioneering work on the red deer, an archetypal polygynous mammal, said that while the genetic findings were intriguing, the next step was to follow up with behavioral observations to determine when females are mating with their preferred males, and how they find them year after year.

"In most studies of harem-forming mammals, where the females don't live in a stable group, we have believed that females did not show strong mate fidelity," said Dr. Clutton-Brock. "Amos's results are interesting because they suggest that females may selectively mate with the same male in successive years. But behavioral observations that they do so are needed to support his interesting genetic results."

Dr. Daryl J. Boness, a research zoologist at the Smithsonian Institution who specializes in the gray seal, said, "The genetic evidence is persuasive, but from my knowledge of gray seals, it's very perplexing."

Gray seals spend most of their time foraging at sea, but each autumn they come ashore to breeding colonies for several weeks. The females give birth to pups conceived the previous season, suckle them to self-sufficient fatness, and then mate again before departing. Most of the females return to the same colony site every year.

The males come ashore for varying times during the fall and spend most of their time fighting; like most male mammals, they have nothing to do with rearing the offspring. A male, at 600 pounds, is about twice as big as the female, a not unusual size differential among pinnipeds.

In performing their field studies, the scientists followed 85 males, most of them clearly dominant, and 88 females for three years. They branded the animals and obtained small skin samples from most of the adults, as well as a total of 120 pups, using the tissue for DNA analysis. They also made extensive observations of how the males distributed themselves around the beach and tallied the number of females they kept within their sight. Yet all their work at maintaining high status reaped few reproductive rewards.

Dr. Amos has no idea why females would bother to seek out the same partner from year to year, when fathers do so little around the house. In most monogamous species, males contribute a great deal to offspring care. He proposes that females do not really like aggressive alpha males, one reason being that a dominant male has to spend so much time battling pretenders to his throne that he may inadvertently crush pups under a flipper.

High pup mortality is a significant hazard in seal society, reaching 60 percent some years. In theory, said Dr. Amos, "partner fidelity should reduce these disturbances and therefore increase pup survival rates." However, he admits he has no data to back up the thesis.

Dr. Boness has another, less romantic suggestion for the widespread fidelity. While the biggest bullies do rule the harems, he pointed out that

North Rona has only a few points of access to the top of the island, where the breeding ground is. "It's possible that males hang around those sites, but that the researchers weren't able to sample these defending males," he said. If so, then the males guarding those access points may exact copulation as a sort of toll for passage. Yet this notion, too, is sheer speculation.

Dr. Boness and his coworkers are now attempting to test the gray seal population on their own study site, Sable Island near Nova Scotia. Tens of thousands of seals congregate there each season, far more than on North Rona, and if females and males manage to reconnect even amid this melee, then the rule book on mammalian mating systems may need to be revised.

—NATALIE ANGIER, July 1995

Deep Underwater, the Breath of Life

Researchers Observe Diving Mammals for Clues to Improving Human Health

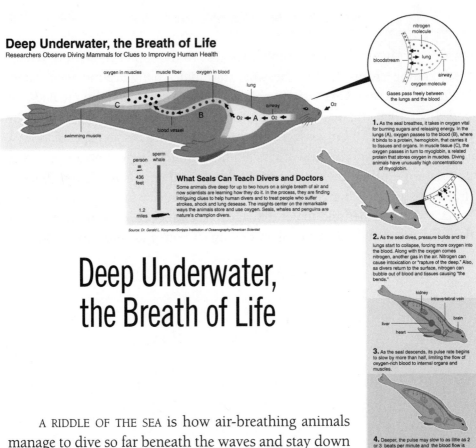

oxygen in muscles · muscle fiber · oxygen in blood

lung · airway · O2

C · B · A · O2 · O2

swimming muscle · blood vessel

nitrogen molecule · bloodstream · lung · airway · oxygen molecule

Gases pass freely between the lungs and the blood

person = sperm whale

436 feet

1.2 miles

What Seals Can Teach Divers and Doctors

Some animals dive deep for up to two hours on a single breath of air and now scientists are learning how they do it. In the process, they are finding intriguing clues to help human divers and to treat people who suffer strokes, shock and lung disease. The insights center on the remarkable ways the animals store and use oxygen. Seals, whales and penguins are nature's champion divers.

Source: Dr. Gerald L. Kooyman/Scripps Institution of Oceanography/American Scientist

1. As the seal breathes, it takes in oxygen vital for burning sugars and releasing energy. In the lungs (A), oxygen passes to the blood (B), where it binds to a protein, hemoglobin, that carries it to tissues and organs. In muscle tissue (C), the oxygen passes in turn to myoglobin, a related protein that stores oxygen in muscles. Diving animals have unusually high concentrations of myoglobin.

2. As the seal dives, pressure builds and its lungs start to collapse, forcing more oxygen into the blood. Along with the oxygen comes nitrogen, another gas in the air. Nitrogen can cause intoxication or "rapture of the deep." Also, as divers return to the surface, nitrogen can bubble out of blood and tissues causing "the bends."

kidney · intravertebral vein · brain · liver · heart

3. As the seal descends, its pulse rate begins to slow by more than half, limiting the flow of oxygen-rich blood to internal organs and muscles.

4. Deeper, the pulse may slow to as little as 2 or 3 beats per minute and the blood flow is directed mainly to the brain and nervous system.

5. Fish and squid, favorite seal prey, tempt the animal into deeper water. Increasing water pressure causes its lungs to collapse, stopping the absorbtion of gases, and preventing a dangerous build-up of nitrogen.

6. The seal can keep swimming deeper, using oxygen stored in its muscle tissues.

7. If the seal stays under long enough, it will use up the oxygen stored in its muscles. It can still dive for a while longer, burning sugars without the aid of oxygen. But this process eventually builds up dangerous acid wastes.

Deep Underwater, the Breath of Life

A RIDDLE OF THE SEA is how air-breathing animals manage to dive so far beneath the waves and stay down for so long. Such acts are obviously no problem for fish, which have gills. But people diving without the aid of special gear can stay under for no more than minutes and go down no more than a few hundred feet before facing blackout and eventual death.

Despite the dangers of probing the icy darkness, some birds and mammals and reptiles do so regularly, plunging to depths of up to a mile or more and staying down for as long as two hours on a single breath.

To scientists, long baffled by such feats, the divers often seemed to break the laws of physiology. Now, however, the secrets of these animals are starting to come to light, and the insights promise to help treat and prevent all kinds of human ills.

N.Y. Times News Service

Gorka Sampedro

Among the discoveries are that the divers manage to survive quite differently from land creatures when holding their breath. They tend to rely far less on air stored in their lungs and far more on oxygen stored in their muscles.

"The field is developing very fast," said Dr. Burney J. Le Boeuf, a seal expert at the University of California at Santa Cruz. "But it's still very difficult."

Dr. Le Boeuf noted that adult female elephant seals, which are among the deep elite, spend 10 months a year at sea and descend so far so frequently to depths where the pressure is crushing that their lungs are judged to be collapsed up to 95 percent of that time.

"Tell that to a medical guy and they probably won't believe it," he said in an interview. "Collapsing is one thing and reinflating is another, and we don't know how they do it."

One emerging clue to the formidable powers of the deep divers centers on their use of oxygen. Their muscles tend to hold unusually high concentrations of myoglobin, a protein that picks up life-giving oxygen from the blood and stores it for later use in providing usable energy for muscles (by oxidizing sugars). Thus, their muscles can work unusually long and hard without requiring the animal to breath fresh air.

Dr. Gerald L. Kooyman and Dr. Paul J. Ponganis, physiologists at the Scripps Institution of Oceanography in San Diego, writing in *The American Scientist,* a bimonthly journal, call the high myoglobin levels of deep divers "perhaps the hallmark characteristic that sets them apart from all land forms." Experts say such physiological tricks can be viewed as a living lesson for the practice of medicine.

Dr. Ponganis, a physician who also has a Ph.D. in physiology, said studies of how the animals thrive at low oxygen levels and at reduced blood flows might one day produce better treatments for shock and stroke victims, who can suffer permanent damage when their brains are deprived of oxygen, as well as new ways to preserve human organs slated for transplantation.

"There's lots of potential applications," Dr. Ponganis said, although he acknowledged that much remains to be learned.

"In the last decade we've made good progress in understanding how deep they go," he said in an interview. "But we know very little of what they do down there. And when you get to the physiology, we know even less."

The feats of the animals seem especially impressive when compared with the modest accomplishments of humans.

Scuba divers with air tanks on their backs can go down only 100 or 200 feet, and then face special hazards. On the way down, nitrogen narcosis, or rapture of the deep, can prompt a kind of giddy stupidity that encourages dangerous risk taking. The drunkenness is thought to be caused by a slowing of nerve impulses and to be akin to the action of nitrous oxide, an anesthetic. On the way back up, the bubbling of nitrogen out of the blood can cause decompression sickness, or the bends, which in severe form can be fatal.

Human divers breathing mixes of special gases can go down deeper still, as much as 600 feet or more. But they can also develop oxygen poisoning, which triggers convulsions, or face the debilitating effects of high pressure, including nervous excitability and fatal seizures.

Even when everything works well, the strains are daunting. Consider the extreme case of Francisco Ferreras-Rodríguez of Cuba, who holds the world record for diving without the aid of breathing gear. In 1996, he plunged 436 feet down. Studies have shown that the sea's pressure at 400 feet compresses his chest size from 50 to 20 inches.

Diving animals go much deeper yet show no signs of the strains and ills that routinely strike human divers. But their skills have evolved over millions of years.

Scientists have only recently begun to realize that many kinds of air-breathing animals dive deep, mainly, it appears, to forage for abundant prey in the icy darkness. Historically, the abyss was considered barren or thinly populated by bizarre fish with eerie lights and sharp fangs. Now it is known to teem with eels, fishes, squids, octopuses, sharks, rays, crabs, tube worms and all manner of gelatinous creatures.

Of the air breathers, the deepest diver of them all is apparently the sperm whale, a toothed beast that probes the deep for giant squid. Scientists have tracked these whales to depths of 1.2 miles, and assume they might go deeper. How they do so is a mystery.

Dr. Kooyman of Scripps said their great size and excitability make them very hard to study. "No one knows much about their physiology," he said in an interview.

Much more is being learned about the diving secrets of the northern elephant seal, which has been tracked to depths just shy of a mile and exceeds the sperm whale in terms of endurance, staying down for as long as two hours.

Studying the elephant bulls is hard because they typically grow to lengths of 16 or 18 feet, weigh up to four tons and throw their weight around, violently at times. But scientists have recently had success in hooking up juvenile seals to small packages of advanced sensors.

In California, seven juveniles were recently wired to study their heart rate. On land it registered a mean of 107 beats per minute. At sea, as the seals dove, it slowed to a mean of 39 beats per minute, a drop of 64 percent, Dr. Le Boeuf of the University of California and his colleagues reported in *The Journal of Experimental Biology*.

The usual pattern was a sharp initial fall in heart rate followed by a gradual decrease throughout the dive. Occasionally, the rate fell as low as 3 beats per minute. In one case, the time between beats was a staggering 26 seconds.

A slow heart hints at an overall metabolic drop, Dr. Le Boeuf said in an interview.

"Clearly, this is very important in explaining how they can parcel out a limited amount of oxygen for so long," he said. "But the great depths of the dives are much more difficult to explain." That, he said, involves surviving the crush of high pressure.

For all deep divers, a key challenge is to fight off decompression illness, or the bends. In theory, the rising pressures that an animal faces while going down should force nitrogen from lung air into the animal's bloodstream and other tissues. Then, as the animal ascends rapidly, the nitrogen bubbles out of solution, wreaking havoc with all kinds of normal physiology, including nerve function and blood flow.

But the animals suffer no such thing. One reason may be glimpsed in the findings of Dr. Konrad J. Falke and his colleagues at Harvard University. They studied Weddell seals, which live in the icy seas around Antarctica and dive to depths of at least a half mile. But no matter how deep the seals dive, nitrogen levels in their blood rise little. That suggests the lungs collapse rapidly, keeping large amounts of nitrogen from entering the blood in the first place. The collapse, after all, halts the flow of all atmospheric gases from the lungs into the bloodstream.

That might seem like a serious liability because it also cuts off the flow of oxygen into the blood. However, research shows that muscles—which burn up oxygen rapidly during deep descents, working much harder than any other organ—carry their own supply in the form of myoglobin, which binds oxygen in the muscles of many animals during normal states of breathing.

Dr. Kooyman and Dr. Ponganis of Scripps say that particularly among deep divers that feed while submerged, levels of myoglobin are 3 to 10 times higher than in land animals.

An example is found in the emperor penguin, which can dive to depths of up to a third of a mile. It stores 47 percent of its overall body oxygen in its muscles, the rest circulating through the blood and the lungs. The figure for the bottlenose dolphin, which also dives up to a third of a mile deep, is 39 percent.

In contrast, humans store only 15 percent of their oxygen in their muscles.

The high storage rate in muscle tissue of deep divers, Dr. Kooyman and Dr. Ponganis say in their article in *The American Scientist,* appears to be an important factor in how the animals can dive so deep.

With muscles highly self-sufficient, the scientists add, oxygen circulating in the bloodstream can be redirected during a dive toward organs that have no storage powers, particularly the brain.

Still other organs, such as the kidney, liver and digestive system, may practically halt operation altogether, Dr. Kooyman and Dr. Ponganis say. For instance, the abdomens of diving penguins have been found to drop markedly in temperature, suggesting a metabolic slowdown.

Dr. Ponganis, the team's physician, is fascinated by the uses that the diving secrets may eventually have for the practice of medicine. He envisions a whole series of medical breakthroughs as new technologies allow scientists to better understand the unique physiological adaptations behind the diving feats.

For instance, he said that the ease with which the lungs of the animal divers collapse and reinflate is probably aided by a special class of chemicals known as surfactants that coat lungs; they also occur in humans. These animal surfactants, he said, might be analyzed and their composition mimicked to help treat human lung disorders.

Another rich area of study is how organs survive despite low levels of oxygen. In humans, kidneys can be hurt quickly if blood flow is reduced, Dr. Ponganis noted. But the diving animals appear unaffected. "The molecular mechanism isn't known," he said. "But if we could figure that out, it could help how we treat patients, or even organ transplantation and heart transplants." Shock and stroke victims might also be aided.

In theory, the information might also eventually be adapted to aid the treatment and prevention of human diving ills, or even to let divers go deeper into the sea, discovering a sunless world that the animal explorers have known for ages.

—William J. Broad, November 1997

Calving of Right Whales Faces New Threats

EARLY SPRINGTIME is the time when female northern right whales and their newborns migrate northward from calving grounds off Florida and Georgia to around Cape Cod, Massachusetts, taking about a month for the journey. The 1996 season has been a good one for baby making among the northern rights. Scientists have sighted 20 calves, a record after years of falling counts. Only 320 or so of the behemoths now ply the North Atlantic, and a high rate of reproduction is seen as critical to the comeback of these big mammals, once hunted to near extinction and now the most endangered of the great whales.

But despite more than a half century of protection, as well as sustained federal and private conservation efforts, the 55-foot, black-and-gray whales are failing to rally and their population remains dangerously low, baffling scientists and alarming environmentalists.

Six whales have died so far this year, including three calves, the highest number of deaths on record for so short a period. Part of the problem is that the lumbering giants swim through one of the nation's busiest sea lanes for commercial shipping and naval maneuvers, at times getting hit. Other whales get entangled in fishing gear.

But scientists say the roots of the problem go beyond such incidents and are increasingly a grim mystery, prompting a redoubling of protective efforts and detective work.

"We don't know what's going on," Dr. Scott D. Kraus, chief scientist at the New England Aquarium in Boston, the main research group working to save the animal, said in an interview. "It gets nerve-racking."

Environmental groups, including Greenpeace and the International
Wildlife Coalition, recently contended that navy war games off Georgia and

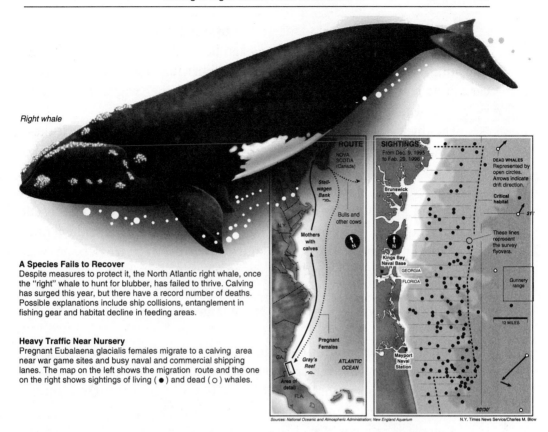

Right whale

A Species Fails to Recover
Despite measures to protect it, the North Atlantic right whale, once the "right" whale to hunt for blubber, has failed to thrive. Calving has surged this year, but there have a record number of deaths. Possible explanations include ship collisions, entanglement in fishing gear and habitat decline in feeding areas.

Heavy Traffic Near Nursery
Pregnant Eubalaena glacialis females migrate to a calving area near war game sites and busy naval and commercial shipping lanes. The map on the left shows the migration route and the one on the right shows sightings of living (●) and dead (○) whales.

ROUTE
NOVA SCOTIA (Canada)
Stellwagen Bank
Bulls and other cows
N.Y.
Mothers with calves
Pregnant Females
GA.
Gray's Reef
Area of detail
FLA.
ATLANTIC OCEAN

SIGHTINGS
From Dec. 9, 1995 to Feb. 29, 1996
Brunswick
Kings Bay Naval Base
GEORGIA
FLORIDA
Mayport Naval Station
DEAD WHALES Represented by open circles. Arrows indicate drift direction.
Critical habitat
These lines represent the survey flyovers.
Gunnery range
12 MILES
31°
80°30'

Sources: National Oceanic and Atmospheric Administration; New England Aquarium   N.Y. Times News Service/Charles M. Blow

Florida with five-inch guns and 500-pound bombs were probably responsible for many of the recent deaths.

But scientists, while happy to question navy practices, often say that what causes the deaths is extremely murky. Right whales dying for any reason, natural or unnatural, float on the surface of the sea, their carcasses vulnerable to damage from passing ships and military maneuvers and often making cause and effect very difficult to disentangle.

"I fear our faddishness," said Dr. Charles Mayo, a senior scientist at the Center for Coastal Studies, a private group in Provincetown, Massachusetts, on Cape Cod, that studies whales. "We're all so desperate to find out what's going on. We have to be careful not to overlook the less conspicuous things. Changes in the coastal ecosystem concern me a great deal—no blood, no carcasses, just silent gasps.

"Something is happening. And given the status of the northern right as severely endangered, everything is guilty until proven innocent."

The past and current dangers confronting *Eubalaena glacialis*, the North Atlantic right whale, stem to a significant degree from its anatomy. The animal, to put it bluntly, is fat, with blubber making up about 40 percent of its body weight, more than virtually any other whale. It is a slow swimmer, seemingly unable to hit speeds over five knots.

The thick layer of blubber keeps the right whale afloat when it dies; most other whales quickly sink. And because of its inherent buoyancy, the whale also tends to rest, feed, court and mate at or near the surface.

For centuries, such attributes made the mammal the "right" whale to hunt and kill—thus its common name. The attraction was mainly its oil, rendered from fat and used as lamp fuel and lubricants and eventually as an ingredient in soaps and paints. The baleen, or whalebone, was strong and flexible and used to make such things as whips and corset stays. Adding to its allure, the right whale tended to dwell near coasts, making it easy prey. It was the first of the large whales to be commercially hunted.

Starting a millennium or so ago, the species was pursued on a large scale, at first by Basque whalers around the Bay of Biscay. As that population of whales withered and knowledge of the world's seas grew, hunting shifted to the western North Atlantic and then the Pacific.

A single whale could yield as much as 90 barrels of oil and 1,200 pounds of baleen. One kill could pay for an entire voyage. Everything else was pure profit.

By 1935, the species had declined to such low numbers that the League of Nations, fearing the whale would become extinct, was able to talk most nations into giving up the hunt.

Since then, the animal's status has remained shaky because of an enigmatic mix of human and natural factors. The main suspects are ship collisions, entanglement in fishing gear and habitat decline in feeding areas. The toothless mammals are skimmers of the sea, feeding mainly on dense swarms of copepods, which are tiny crustaceans the size of a match head.

Federal and private efforts to save the whale increased in the 1970s and 1980s and generally focused on trying to achieve a better understanding of the animal's habitat and habits. Dr. Kraus of the New England

Aquarium pioneered surveys at sea and learned to tell individuals apart, creating a catalogue of photographs. Particular right whales bear distinctive calluses of hardened skin, as well as characteristic scars and coloration.

It turned out that the main surviving herd of northern right whales migrates along the East Coast. In summer, the animals frequent the Bay of Fundy, moving south around Cape Cod in the fall and winter. Pregnant females, Dr. Kraus and his colleagues discovered, travel in the winter to warm, shallow areas off Georgia and northern Florida to give birth to their young. In spring, the migration route is reversed.

Babies are up to 15 feet long at birth and weigh almost a ton. Mothers nurse the calves for about a year. This long period of lactation is hard on the nursing females, who require one to three years to recover from a birth. Gestation itself takes about one year.

Alarmingly, the aerial surveys of the whales found that the double-digit births of the 1980s were giving way to a trend to fewer babies—17 in 1991, 12 in 1992, 6 in 1993, 8 in 1994 and 7 in 1995.

Amid that drop, the federal agency that protects endangered sea mammals, the National Marine Fisheries Service of the National Oceanic and Atmospheric Administration, declared an area off the Georgia and north Florida coast as a critical right-whale habitat.

Aerial surveys of the area were stepped up, as were warnings of whale movements to mariners in the hopes of avoiding collisions. And the federal budget for right-whale research nearly quadrupled, going from $220,000 in 1992 to $850,000 this fiscal year, aided by the Clinton Administration's general support of ecological studies.

This winter, when right-whale births and deaths both unexpectedly soared, the latter events set off alarm bells throughout Washington.

"We usually get one or two deaths a year," said Michael Payne, a right-whale protection expert at the National Marine Fisheries Service. "Now we have six, and the year just started. That's a problem—big time."

One of the recent deaths, that of a 44-foot adult, occurred at Cape Cod. More ominously, the other five were near the southern calving grounds. Of the three calves that died, one apparently had genetic defects, one had no signs of trauma other than congested lungs and one had hemorrhaging behind one eyeball.

"These changes are not inconsistent with a concussion event," Dr. Gregory D. Bossart, an animal pathologist at the University of Miami, said of the hemorrhaging in a report to the national fisheries service.

Criss-crossing the calving ground are ships from large commercial ports like Brunswick, Georgia, and Jacksonville, Florida, as well as from major navy installations like Mayport in Florida and Kings Bay in Georgia, the East Coast home of the Trident submarines. Military traffic in the region is estimated at about one-tenth of the total.

The unusual spate of deaths occurred around February, when the navy was conducting firing exercises and fielding a large number of ships. In early March, officials from the navy and the National Marine Fisheries Service began meeting three times a week in Washington to discuss the issue and work out new precautions. It was the first such federal meeting over right whales.

Outside groups quickly joined the debate. "Five endangered whale deaths linked to U.S. Navy," a Greenpeace news release charged on March 13, 1996.

Twenty-one members of Congress, all Democrats except for one Republican, wrote Defense Secretary William J. Perry on March 19 to lobby for stepped-up safeguards, including extended aerial surveys. "Special caution must be exercised by all parties," they said, "given the extremely precarious nature of this species."

The navy denies responsibility, saying the evidence is ambiguous and that the only dead calf with signs of trauma was found 50 miles from the exercises. Even so, the service is taking steps to move its war games farther from the calving ground.

"We don't have a clue to what killed those calves," Comdr. Stephen Pietropaoli, a navy spokesman at the Pentagon, said in an interview. "We're very interested in doing whatever we can to determine the cause of these deaths and to take whatever additional steps we can to reduce what we think is a very small hazard posed by navy operations."

The service is forming a blue-ribbon panel to review the necropsies and is paying for an extension of the aerial surveys off Florida and Georgia until biologists judge that the new calves and mothers have migrated substantially to the north. The surveys are conducted by the New England Aquarium in small planes.

In addition, navy aircraft and helicopter pilots in the area have been ordered to search for right whales and report their locations as an aid to avoiding collisions.

"The navy is committed to serving as a responsible steward of the marine environment and marine life," Steven S. Honigman, the navy's general counsel, replied to the congressmembers in a letter dated March 22.

The Justice Department is involved in meetings on the death issue, since some federal officials worry that private groups might sue the government for negligence under the Endangered Species Act. Such lawsuits have been filed dozens of times over the decades.

Attention is now shifting to long-term protection efforts. One emerging issue is that the females apparently travel more extensively while giving birth than originally thought, with whales increasingly sighted outside the critical habitat. Federal officials are now considering an expansion of the zone.

"It's become clear that the whales move freely beyond that area," said Dr. Chris Slay, director of aerial surveys and southeastern research for the New England Aquarium, who is based in Fernandina Beach, Florida.

Experts caution that safeguards and collision avoidance can only do so much and that some factors may be beyond easy control. Some preliminary research suggests that the whale's storehouse of fat also stores an array of toxins, possibly affecting the animal's overall health.

Experts also worry that inbreeding among the small population of remaining whales is reducing their vigor and making them more vulnerable to birth and genetic defects.

For the moment, the surviving calves from this birthing season are to be monitored very carefully as they migrate northward with their mothers. With luck, scientists say, new clues will eventually emerge in the riddle of how to save the rarest of the great whales.

"This is the only whale species we might lose in our lifetimes," said Dr. Kraus of the New England Aquarium. "They're extremely vulnerable to things we don't understand. And they're really dumb about ships. That's the one thing we understand and can do something about."

—WILLIAM J. BROAD, April 1996

6

THE PRIMACY
OF PRIMATES

Primates are mammals that have left the ground but not attained the air: Their bodies are designed for living in trees.

But just as some mammals have returned to the sea, so some primates have returned to the ground, while retaining the overall features of their tree-living existence. Chimpanzees walk on their hind feet and knuckles; humans still have the long arms of their branch-swinging forebears but have refashioned the function of their hind feet from grasping to walking.

Primates fall into two main groups, called prosimians and anthropoids. Prosimians, as their name implies, are not quite monkeys. Though they have bodies and tails like monkeys, their faces are more snoutlike. Prosimians include the lorises, lemurs, tarsiers and the aye-aye; the anthropoids are the monkeys, apes and humans.

Anatomists have difficulty in defining unique primate features because primates, by and large, are generalists in terms of their body plan. They tend to have long tails for balance, and hands and feet suited for grasping branches. They have a sharpened sense of sight, perhaps to help judge distances accurately as they navigate through the forest canopy, and a duller sense of smell, an ability that is less important to tree dwellers.

They have larger brains and a more developed cerebral cortex than other mammals. The larger brains, at least in relation to their body size, are part of what is a distinctive feature of primates—their behavior. As the following articles describe, the social interactions and daily routine of higher primates are considerably richer than those of most other animals.

Bizarre Baby Raises Hopes for an Endangered Primate

THE AYE-AYE IS A creature that can be described only by comparing it piecemeal with other things. It is the size of a cat, has the ears of a bat, the snout of a rat, a tail like a witch's broom and a long knobby middle finger that would look just fine on that same witch's hand. Its teeth are as tough as a beaver's and its eyes bulge out from its skull like a tree frog's. And when a baby aye-aye cries, it sounds like a squeezed rubber duck—particularly when it is being held with exceptional clumsiness by a visitor attempting neither to hurt the little beast nor to be hurt herself.

This is, after all, the first aye-aye to be born in captivity outside Madagascar, its native home, and its birth signals a possible turnaround in the fate of the animal considered the world's most endangered primate. Thus, it is an honor to be allowed to pick up the three-week-old baby and feel its coarse fur, its writhing, protesting muscles, its tiny heart thumping in fear and fury.

At the same time, the researchers at the Duke University Primate Center, where the aye-aye was born early in April, have spoken graphically of the baby's efforts to bite its handlers, and of how an aye-aye's teeth can pop the top off a coconut in moments. As the 12-ounce creature eeps shrilly and squirms its head this way and that, one wonders if the little darling would be injured too badly were it to be dropped on the floor.

"Well, I'd say Blue Devil is ready to return to his mother," said Dr. Elwyn L. Simons, a primatologist who runs the center at Duke and who named the baby after the university's championship basketball team. "I think he's had enough excitement for one day." Yes, please.

Dr. Simon is a rotund man who speaks slowly, pads about as quietly as a panther and can imitate with delightful precision the gestures of the

Night Prowler of Madagascar
The aye-aye is remarkably
adapted to fill an ecological
niche occupied elsewhere by
woodpeckers. When its ex-
traordinarily long, bony
middle finger taps a tree,
its batlike ears can sound
out any insect grubs in the
hollows within. Then, with
its chisel-shaped front
teeth, it rips through the trunk
to feed. The finger is a multi-
purpose tool, for poking holes
to extract liquid from eggs or
coconuts. Aye-ayes build big
sleeping nests in the forks of
trees and share both nests and
food. Their nocturnal habits
and menacing digits have
made them objects of super-
stition in their homeland, helping push them near extinction.

Dimitry Schidlovsky

endangered primates he cares for. "Watch this, this is how an aye-aye drinks," he said, and then jerked his middle finger rapidly back and forth, back and forth from an imaginary source of liquid and up to his mouth, slurping ever so slightly for effect. Later, when he bit into a banana, he grinned slyly, fully aware of the portrait he presented as he ate.

More often, however, Dr. Simon appears quite serious, and with reason. He and his colleagues at the primate center, as well as other scientists at zoos and universities, are struggling to save the aye-aye and other lemurs from extinction. The 30 species of lemurs alive today are limited almost exclusively to Madagascar, an island off the coast of eastern Africa that is one and a half times the size of California. And because the forests of Madagascar are dwindling to the vanishing point, as an impoverished and rapidly swelling human population slashes and burns the forests simply to survive, all 30 species are considered endangered.

Dr. Simons and his colleagues are trying to breed the lemurs in captivity, with the hope of eventually reintroducing at least some of the species back into national reserves on Madagascar. They are also striving to learn everything possible about lemur desires, habits and appetites, and any other insights that can be used to better the animals' chances for survival in the wild. As a group, lemurs have been much less intensively studied than, for example, chimpanzees and orangutans.

A Yale University doctoral student recently completed the first long-term study of aye-ayes on Madagascar, and even after 18 months of extremely difficult research she has only just begun to understand their social structure, courtship rituals and other basic questions.

Although lemurs are prosimians, or premonkeys, and in brain size and other features are more primitive than the so-called higher primates, they are extraordinarily vivid, some with faces like quizzical little monks, others with the shocking blue eyes of Paul Newman. The tiniest species, the mouse lemur, is six inches long and is the world's smallest primate, while the biggest lemur, the indri, is almost three feet long.

The primate center has been extremely successful at rearing lemurs, and it now has more than 400 representatives of fifteen species scampering about in large cages or through 65 acres of open-air enclosures in the North Carolina woods. Researchers from the United States and Europe come to the center to study lemur behavior as the primates squabble, forage, mate, raise their young, groom one another with their comblike teeth and otherwise carry on in conditions that approximate wilderness, although the outdoor enclosures are fenced with wires that will deliver a very mild shock should a lemur develop undue wanderlust.

Some primatologists are particularly interested in the lemur as a kind of living fossil, a creature that survived through the lucky accident of its geographic isolation on Madagascar. Elsewhere lemurs became extinct, displaced by the bigger and craftier monkeys and apes; but some prosimians migrated from the African mainland to Madagascar about 50 million years ago, perhaps by floating on vegetation, and thereafter they flourished without the pressure from higher primates or indeed from any significant predators.

"They evolved like Darwin's finches," the birds that live on the Galápagos Islands, said Andrea Katz, a research scientist at the primate center. "Each one fills a different niche." Researchers believe the prosimians hold clues to the evolution of social behavior among ancestral primates.

Others are intrigued by the lemurs' reversal of standard sex roles. Among the so-called higher primates, males are often larger than females and thus frequently bully them. By contrast, male and female lemurs are similarly sized, and the female dominates the male, eliciting from him displays of submissive behavior and shooing him away whenever she grows annoyed.

At the Duke facility, Dr. Simons and his coworkers have managed to breed several species of lemur that others had found impossible to rear in captivity, including the golden-crowned sifaka, a slender blond acrobat that leaps from one branch to another in strange sideways arcs. The researchers succeeded by coddling the reluctant primates 24 hours a day and supplementing their diet with exotic treats like mango leaves from Florida.

"You have to take as much care of them as you would your own children," he said.

Perhaps no challenge will be as great as the one he has undertaken to help the aye-aye. Not only is the aye-aye suffering from an ongoing loss of habitat, as are all the lemurs; but it has another problem that makes its especially vulnerable. The people of Madagascar, the Malagasy, either ignore or respect most of their lemurs, calling them "the little men of the forest." The aye-aye is an exception: It is considered an evil omen, a harbinger of death. By one legend, should an aye-aye point its elongated middle finger at you, you are destined to die, swiftly and horribly.

To rid themselves of the curse, many Malagasy will kill any aye-aye they see and then place its corpse on a stake in a crossroads, with the hope that a stranger will pass by and absorb the aye-aye's malevolence.

The taboos surrounding the aye-ayes are so pervasive that some think the primate, whose scientific name is *Daubentonia madagascariensis,* gained its common name as a spinoff of the Malagasy expression for "I don't know," suggesting that even to mention the creature is forbidden.

While traveling through the countryside in search of aye-ayes, Dr. Simons often heard accounts that the villagers had killed five or more animals just before his arrival.

"In the villages they're treated with the same alarm or disgust that people here express when they encounter rattlesnakes," said Dr. Simons. "They ask, 'Why on Earth would you want to capture those things and take them to another country?' "

He said he suspected that one reason the animal is hated is its bizarre appearance. The aye-aye doesn't look like any other primate on the planet, and in fact it was first classified by French researchers in the 18th century as a squirrel. Its long fur is a dusky, forbidding shade of black, and in the dark its yellow eyes gleam demonically. The animal also has the dangerous habit of being curious about humans, making it an easy target for those who want to kill it.

Dr. Simons said one reason the aye-aye has not been exterminated is that it is nocturnal. Most Malagasy villages lack electricity, so the people generally retire to their homes after sunset, shortly before the aye-ayes begin foraging.

But the animal has its appeal. Its brain is larger and more deeply convoluted than that of any other prosimian, suggesting a somewhat greater intelligence. Its hearing is so keen that it can tap on a tree trunk and detect the hollow regions within, indicating the presence of the beetle grubs it covets. The animal will then rip through the trunk with four chisel-shaped front teeth that, unlike those of other primates, will grow throughout life. And of course there is the aye-aye's extraordinary middle finger, a long thin digit that can bend in every direction, even backward to touch its forearm. The finger is an all-purpose tool for delicately tapping the tree trunks, poking holes in eggs, pumping the liquid out of those eggs, and extracting milk from coconuts.

Given that aye-ayes are night creatures and that Madagascar's long, sodden rainy season discourages most researchers, the animal has scarcely been studied. But Eleanor J. Sterling, who is doing her dissertation on the aye-aye, recently spent 18 months following the creatures on an uninhabited island off the northeast coast of Madagascar, tagging the primates with radio collars and tracking them from dusk to dawn, her effort illuminated by a headlamp.

She learned that the animals, long thought to be solitary, in fact spend a considerable amount of time socializing. They build huge sleeping nests in the forks of trees, and they willingly trade nests from one night to the next, with the animals seeming fairly communistic about their property. Their diet proved richer in fruits and vegetables than anybody had suspected, and they seem able to breed several times throughout the year, rather than only during a single season, as other lemurs do.

She also discovered that aye-ayes adopt a kind of Kamasutra approach to lovemaking. When a female decides she is ready to act, she will hang upside down from a branch, and a male will position himself by entwining his legs around her ankles, himself facing downward, and then grasping her about the chest, with the weight of both supported by the female. The pair will then copulate for an hour or two, much longer than the usual primate session.

"In the meantime, a bunch of other males will be climbing up the tree, trying to get him off her and mate with her as well," said Ms. Sterling. "The whole thing is a three-ring circus." The female may eventually mate with more than one partner before her estrus is through.

She did not determine how long the gestation period is, although it is likely to be at the upper end of lemur pregnancies, around 140 days.

Dr. Simons said he believed that the newborn Blue Devil was conceived during just such gymnastics in the forests of Madagascar last November, only days before the infant's mother, Endora, was captured by the Simons team and taken back to Duke along with three other aye-ayes. There they joined three already at the center.

Dr. Simons has found that, contrary to long-standing claims that the creature is limited to patches of forest along the coast of Madagascar and thus it is too rare to risk taking any from the wild, the aye-aye also lives in woods farther inland.

The primate center is optimistic that another of its females may have recently conceived. Eventually, if the aye-ayes reproduce well in captivity, the primate center may distribute several to zoos around the country. The only zoo in the Western Hemisphere with aye-ayes is in Paris.

It will be more difficult to devise a long-term strategy for the primates on Madagascar. Since the first Indonesian settlers arrived on the island 1,500 years ago, about 85 percent of its spectacular tropical forest has been slashed and burned by humans for wood and farm and grazing land. The island's thin soil is badly eroded and its nutrients depleted, further threatening the forests that remain; and the human population continues to grow at 2.1 percent a year, one of the fastest rates in the world.

Yet scientists say there is an enormous will now to rescue Madagascar, which is plush with thousands of species found nowhere else, including 142 species of frogs, 106 types of birds, 6,000 species of flowering plants, and half the world's chameleons. Recognizing the potential of its wildlife as a source of income through, for example, "ecotourism," the government has opened its doors to researchers, conservation groups and other international efforts.

But whether the wealth of species can be preserved while so many Malagasy remain impoverished, and whether the people can ever learn to view the aye-aye as lovable rather than malicious, remain frighteningly open questions.

—NATALIE ANGIER, May 1992

Cotton-Top Tamarins: Cooperative, Pacifist and Close to Extinct

COTTON-TOP TAMARINS are the punks among monkeys: small in stature, big in hair and noisy. At six inches high and weighing barely a pound, they are one of the world's tiniest primates. A shock of white fur rises from the midline of the skull in a kind of Einsteinian Mohawk, dipping down toward ears so prominent they beg to be adorned by a row of hoop earrings. And the monkeys make gorgeous punk music, squealing, whistling, chirping, letting loose with slicing screams.

But lest their singing be dismissed as so much senseless muddle, researchers say the tamarins' vocal repertory of 38 distinct sounds is unusually sophisticated, even by primate standards. It is structured like a language, conforming to grammatical rules and able to express curiosity, fear, dismay, playfulness, warnings of predators approaching, a desire to be groomed, joy over food discovered and calls to the young among them to join the feast.

For if there is one thing that is not punk about these creatures it is their attitude. No in-your-face surliness or social anomie here. Cotton-tops are cooperative and pacifist to a surprising degree. Adults share food with the young in their group, even those that are not related to them. They carry one another's children around and protect them from danger. They accept newcomers into their fold so readily that researchers who study the monkeys in their native Colombia have not been able to detect a single behavioral difference between longtime residents of a tamarin troop and the monkeys who pop in one day and make themselves at home.

"They act as if they lived there all their lives," said Dr. Ann Savage, a research scientist with the Roger Williams Park Zoo in Providence, Rhode Island, considered one of the best zoos in the nation for its extensive conservation efforts. "They immediately start doing sentry duty and caring for the young."

Now, Dr. Savage, along with Dr. Charles T. Snowdon and Dr. Toni E. Ziegler of the University of Wisconsin in Madison, have discovered new evidence of just how much cotton-tops dislike stress and conflict. Working with both wild populations and colonies kept in captivity, the scientists have been studying the cotton-tops' extraordinary system of reproduction and child care, in which adult monkeys voluntarily forgo their own fertility for long stretches of time while they practice being good parents by rearing other monkeys' offspring.

At a recent meeting of the Animal Behavior Society held in Seattle, the scientists presented surprising results on the hormonal mechanisms through which nonbreeding females keep their own reproduction in check. They had expected to find evidence that the breeding female kept the other females docile through intimidation and constant stress, but instead they detected clues that the helping females in fact were utterly unfrazzled and seemed to be choosing not to ovulate until their turn had come.

The latest work offers a counterpoint to the familiar face of nature, that of animal brutality, hypercompetitiveness, selfishness and deceit. Indeed, a rising number of animal behaviorists believe that scientists have focused too narrowly on the violent and competitive aspects of the natural world, while neglecting to consider the role of cooperation, friendship and peacemaking in the everyday affairs of many social animals.

Dr. Frans de Waal, a research professor at the Yerkes Regional Primate Research Center in Atlanta and the foremost advocate of this more encompassing view of nature, has argued that for every example of animal ugliness, such as male lions systematically butchering the cubs sired by their competitors, there are cases like the cotton-tops, which live in a primate version of a kibbutz. Yet when Dr. de Waal talks about things like love in animal societies, "some evolutionary biologists dismiss these views as excessively romantic," he said in an interview.

Beyond the inherent interest in understanding the biochemical and behavioral details of how another species manages its social life, Dr. Savage and her peers are desperate to understand cotton-top breeding in order to help the animals survive. Right now, the tamarins are one of the most endangered primates in the New World, if not the entire world. There are fewer than 2,000 of them left in the wild, living in the patchwork quilt of tropical forests that remain in northwestern Colombia. As everywhere in the Tropics, the forests are dwindling fast. According to the books, said Dr. Savage, there are supposed to be about 40,000 acres of forest left in the northwestern region, but surveys indicate the true figure is one-third of that.

Apart from the loss of its habitat, the cotton-top has also suffered from one unique aspect of its physiology. It is the only species, apart from humans, that spontaneously develops colon cancer. As a result, research laboratories in the United States in the past imported them with the hopes of better understanding the human disease. In the 1960s and 1970s, 20,000 to 30,000 cotton-tops were imported from Colombia.

Right now, the monkeys' status as an endangered species prevents any further importation of wild cotton-tops, and only 1,150 monkeys remain in medical laboratories. (Another 700 or so live in zoos or private institutions.) However, the New England Primate Research Center has petitioned the United States Fish and Wildlife Service to change the status of captive cotton-tops from endangered to threatened, which would make it easier to carry out research on the animals without going through excessive red tape. Dr. Savage and her colleagues are fighting against the change in status, arguing that it would make it much harder to prevent the illegal importation of cotton-tops from the wild.

Dr. Savage, who is 33 years old, six feet tall and has a rich, ready laugh and the profile of a Modigliani woman, is doing everything in her power to rescue the cotton-top. She regards it as an excellent indicator species, an animal far enough up the food chain that if it is doing well, so must its habitat. She recognizes its "high cuteness factor," she said, and the natural affinity that people feel for a species that places emphasis on good parenting skills. "People get a warm feeling about them and want to save them," she said.

The cotton-top has also proved to be a tremendous teaching tool. Working part of the year in Coloso, a village not far from the research station in Colombia where the scientists study the tamarin, Dr. Savage has gone into local schools and talked up the tamarin. She found that most of the students had never ventured into the forest, though it was only a couple of miles from their village; and almost none knew either that the cotton-top was so endangered or that it was exclusive to their country. "They thought of it as we might think of a squirrel," she said. "They had no idea it was unique to their backyard."

With a frayed shoelace of a budget, she has gotten the local students involved in research projects on the tamarin and other elements of their local habitat, and she has fostered collaborative enterprises between junior high and high school students in Colombia and New England, conducted via fax machine. Murals, posters, T-shirts, stuffed toy tamarins—anything and everything has been used to promote the monkey as a source of local pride. Political conditions sometimes make their work extremely difficult. Dr. Savage was once held at gunpoint by guerrillas, who thought she was some sort of government agent. But after she showed them a video of the work she was doing, the guerrillas released her.

"I can't imagine my life without cotton-tops," she said. "What would happen if I couldn't go back to Colombia? That would be heartbreaking. Going into the forest and seeing a baby tamarin up in the trees is like the feeling some people get when they go into a great church."

The temple of the tamarin is one that few scientists have entered. When the researchers began studying the Colombian monkey in 1988, they were only the second team to investigate the animals in their native setting. They focused on several populations totaling about 100 monkeys. Because the animals are tiny and live about 15 feet up in the canopy, they are difficult to detect and even harder to distinguish as individuals. To ease their efforts, the scientists captured the monkeys and attached small radio transmitters to several of the males; they also dyed each monkey's white tuft a shade of magenta or yellow or green, which made them look punkier than ever.

More recently, the scientists have begun mastering the art of finding the dime-size drops of feces that the monkeys deposit in the leaf litter, and they are learning to isolate hormones from the droppings to tell the emotional and reproductive state of the various members of a group. In the past, scientists could only perform such tests with blood samples, which are impossible to get on a regular basis and skew the results by putting the animals under stress.

In the emerging portrait of cotton-top life, the scientists have determined that the monkeys are so-called cooperative breeders: A single male and a single female do all the breeding, while the other monkeys in the group forgo their own reproduction to help with the royal couple's offspring. The mother needs all the help she can get to rear young. She gives birth each year to a set of twins that together total up to 25 percent of her body weight, the equivalent of a 130-pound woman giving birth to twins weighing 16 pounds apiece. And then the huge babies must be nursed.

"The mothers have very heavy demand on their system," Dr. Snowdon said in an interview. "This wouldn't work unless there were helpers available." In a colony, which numbers about six, everybody is a helper, carrying the young on their backs, protecting them from predators, feeding them once they have been weaned and engaging them in play. And unlike many other cooperatively breeding mammals and birds, cotton-tops do not seem to be fixated on kinship. By standard evolutionary thinking, cooperative breeders willingly forgo reproduction to help raise their siblings because in so doing they are ensuring the survival of their own genetic lineage. But cotton-top strangers also take part in child care.

The scientists propose that this practice may give them the benefit of experience. Through studies of zoo-living cotton-tops, the scientists have determined that a new cotton-top mother who has had no training in child rearing before becoming pregnant will end up either killing, abusing or fatally dropping her young. For their part, males who act the part of the caring baby-sitter may be showing off to females, particularly when the male has migrated to a new group and hopes to become the breeding father soon. "It's been shown that males carrying infants are more able to copulate with females than those that weren't," said Dr. Snowdon.

Most recently, the scientists have tried to determine just how it is that the nonbreeding females in a group are kept from ovulating, and from trying to have babies of their own. In other cooperatively breeding animals, like dwarf mongooses, stress is thought to play a role. The dominant female bullies her subordinates, cowing them into a hormonal state of infertility. In the case of the cotton-tops, the scientists almost never observed instances of aggressive behavior by the great queen mother, but they nevertheless thought the victimization could be subtle.

In the data presented at the Animal Behavior meeting, however, Dr. Snowdon said they could find absolutely no evidence of elevated stress hormones like cortisol in the nonbreeding helper females. In fact, he has proposed that a lack of stress keeps them relatively subdued. Their studies with captive monkeys indicate that stress hormones begin to shoot up once a female has been taken out of her home group and placed with a new male. Her juices flowing, the female begins to ovulate shortly afterward.

Dr. Snowdon proposes that the dominant cotton-top female may not be keeping her minions in check so much as they are choosing to relax and wait. "Part of the control of reproduction may reside with the subordinate female," he said. "It may be advantageous for her to regulate her own fertility and essentially inhibit herself, until the time when conditions are best for her."

In the wild, the female may need to bide her time for years before she locates a vacancy as colony mother. During the six years that the researchers have been observing the monkeys in Colombia, they have witnessed only one confirmed case of a female who managed to fill a maternal niche away from her birthplace, though they have hints that two other females may also have begun breeding far from home. The tamarins may not feel any great pressure to go forth and multiply. They are a long-lived species, often surviving well into their twenties; and given the group's coordinated efforts to watch for dangers, they only rarely end up as a predator's meal.

Human encroachment, however, is another matter. Indeed, as much as scientists have come to respect the tamarins' courteous and deliberate ways, they may wish there were a few upstart females willing to buck the rules and start breeding a bit before their time—and before the time for the cotton-top tamarin is up for good.

—NATALIE ANGIER, September 1994

Status Isn't Everything, at Least for Monkeys

WHAT CORPORATE EXECUTIVE or member of the English aristocracy would not delight in subordinates like these? Pliant, respectful, uncomplaining, with nary a thought of stealing your job or toppling your privilege! Hardworking, with solid family values, and just begging for the opportunity to attend to your personal grooming. It is so difficult to find good help these days, and harder still to find an employee who will happily fetch your coffee and dry cleaning.

Would that it were possible to hire a coterie of low-ranking rhesus macaques. They would have the fleas plucked from your fur as fast as you could say "top banana."

Rhesus monkeys, those fidgety, expressive, toothy natives of India, are the quintessential social primates, and for "social" read "hierarchical." They live in groups of several dozen animals, each ranked as though a little number were painted on its taupe fur. Dominant monkeys can displace subordinates from choice spots or nutritious pickings, and the offspring of dominant mothers can easily push aside lesser adults many times their size. Biologists have known about rhesus rankings for years; in truth, they are hard to miss.

Yet there may be more to the structure of these dominance hierarchies than meets the eye—or less. To American observers reared on the ideology of polarity and the zero-sum game, of winners and losers, who's hot and who's history, rhesus monkeys appear to be engaged in an obsessive and never-ending struggle to rise to the throne, or to stay there once elevated. The alpha monkeys seem to have the underling members cowed, while the subordinate animals surely must resent their inferior status and be ever on the alert for opportunities to mutiny.

After all, dominant status looks like the winning number in the Darwinian raffle, the ticket to reproductive success, more and healthier offspring, safer territory, the complete caboodle.

Lately, however, some scientists have begun to question the notion that dominant status is the ultimate goal of a social animal, or that those who are subordinate resent their status and wish to move up in the world. Nor does a dominant position guarantee that one will have more or better babies, as had always been supposed. Using DNA fingerprinting to identify paternity of offspring, or following groups of animals for a sufficiently long time to determine which mothers breed most often, scientists have found a number of examples where high status does not translate into reproductive primacy. In fact, animals slightly lower on the social pyramid often have distinct advantages over the alpha animals, if for no other reason than because they need not waste time defending their status and can instead focus on love and family.

Moreover, some animals appear to do best in attracting mates when they cultivate an image, not of a driven winner, but of a softie, an easygoing friend, one who would rather spend time engaged in mutual grooming than in strutting around with chest expanded or claws extended. This tactic of amiability can work for both females and males, giving the lie to the common assumption about where in nature's lineup the nice ones finish.

Scientists also realize that they must pay more attention to the individual personalities of the social animals they study, to avoid classifying an animal as simply dominant or subordinate and instead ask, Is this creature even-tempered and cool in battle, able to assess threats and avoid them when possible? Or is the animal a hair trigger—aggressive or high-strung to a counterproductive degree? Studying the physiology and stress responses of baboons, Dr. Robert M. Sapolsky, a primatologist at Stanford University, has learned that there are at least two distinct flavors of dominant animals, one with a sufficiently stable personality to benefit from social success for many years, breeding all the while, and another, edgier, stress-prone type that burns out quickly, falls from grace and in the end has spawned fewer young than low-key counterparts who never sought prominence in the first place.

In sum, the emerging portrait of the group life of many species is less of a monkey-beat-monkey world of restless status seekers and more of a complex society in which a variety of scripts get played out depending on environmental and temperamental circumstances. Sometimes the ones who are the most obsessed with determining the dominance ranking of a social species are the scientists doing the observing.

"The image we've had is that the rhesus monkeys must spend a lot of time establishing and maintaining rank," said Dr. Kim Wallen of the Yerkes Regional Primate Research Center, one of the foremost facilities for studying primate behavior and physiology. "But the group may be much less stratified than we've tended to see it. Yes, we can recognize ranks, but what your rank is may not have that many consequences. What really may matter is whether you're a member of that group or not."

Scientists warn, however, that the recent questioning of the importance of social dominance should not be taken too far, and that to say high status is not the whole story does not mean it isn't part of the plot.

"The fact that we don't always see a correlation between dominance and reproductive success leads some to say dominance isn't important, and I think it would be a mistake to go to that other extreme," said Dr. Barbara Smuts, a primatologist at the University of Michigan. "I view the progress of science as being the slow erosion of the tendency to dichotomize."

The new view could be significant for understanding the evolution of human social behavior, wherein a concern with rank and achievement are viewed as natural, while gentleness is considered either a denial of darker impulses or a cowardly refusal to get in the ring with fists deployed. In fact, amiability or relative noncompetitiveness in humans' evolutionary past may have proved quite canny reproductive strategies, with as much to recommend them as power grabbing.

Rhesus monkeys do not seem at first glance to be good candidates for the study of why dominance is overrated. The monkeys live in female-centered social groups with alpha females and their kin making it quite clear on a regular basis that they reign supreme. If a subordinate monkey annoys a dominant monkey, the alpha animal and her entire family will punish the offender, backing it into a corner, screaming, biting, swiping and chasing. Hearing the melee, the relatives of the inferior animal do nothing, unwilling to help anyone foolhardy enough to quarrel with a superior. Usually, the subordinate animal attempts to make amends afterward by offering the alpha female a lengthy grooming session.

It was through observing the sexual politics of rhesus life that Dr. Wallen began to doubt the standard interpretation of hierarchy. A big, bearded, ursine man who displays little of the type-alpha sensibility, he noticed that the rank of a female bore no relation to her capacity to fulfill an essential social role: bringing a new male into the troop. Rhesus females control the ability of strange males to gain entry into their tightly guarded social unit. If a female "sponsors" a supplicating male, he is allowed to stay and become part of the family; if no female shows interest, his greater size or canine length count for nothing. The females can and will band together to chase him away.

As it turns out, subordinate females sponsor as many debutant males as do the alpha females. And once the males are in, the sexual activity becomes as freewheeling as a Dionysian holiday. Ovulating females regardless of rank often mate with most if not all of the half dozen or so males in their troop of 30. For their part, although the males establish a hierarchy among themselves, the alpha male does not seem able to monopolize estrus females, nor does he seem particularly concerned about doing so. As a result of the rampant promiscuity, the females almost always get pregnant during their fertile season, and all the males have excellent odds of fathering a few of the offspring.

Dr. Wallen proposes that the most important ranking process for the males occurs before they enter the group, when they are wandering the forest or plains as young bachelors looking for a band of females to give them shelter. At that point, the males may fight viciously with one another, and the weaker or less crafty ones may fade into oblivion. But once a male has been endorsed by a rhesus female and given sanctuary, said Dr. Wallen, the females consider him as good as the other resident males.

In other words, even though the males and females each form their gender-specific hierarchies within the group, dominance does not correlate well with reproductive output, or indeed any other measure of so-called fitness.

"We've focused so much on the fact that the group is hierarchical that we've assumed there must be a benefit to being the highest ranker, and everybody must be striving to be the dominant animal because it brings the best of everything," said Dr. Wallen. "But I don't think that's holding up. I don't think it's holding up in terms of reproductive success," or ability to get significantly more food, he added.

If the rhesus caste system does not exist because the ruling class uses it to hoard all the treasures, why, then, have a hierarchy in the first place? Dr. Wallen proposes it arose from the principle that any social order is better than no order at all.

"Structure produces predictability," he said. "It means you spend very little time figuring out what your social relationships are, and you can focus more on things like mating, foraging, watching out for predators.

"If that's true," he added, "then the model we have of low-ranking animals striving to be high-ranking animals probably really isn't accurate. The low-ranking animals may be perfectly happy as long as they're getting mating opportunities and as long as they're getting fed."

The subordinate animals may even cultivate an image of inferiority simply to keep the social peace. In studies at Yerkes with his graduate student Christine Dray, Dr. Wallen observed that when low-ranking monkeys were in the presence of their betters, they appeared incapable of learning to find a series of hidden peanuts. When the dominant animals were removed from the pen, however, the subordinates headed right for the concealed treats, demonstrating that they had been playing dumb all along.

In other primate species, like gorillas, the struggle for social ascendancy can result in reproductive triumph, but it appears that the best spot for a male is not at the top but one or two tiers down. Recent data on chimpanzees living in the Gombe Stream Research Center in Tanzania indicate that the rising males in a group, not the top male, sired the most offspring. "The alpha male is like the dean or the chairman of a department," said Dr. David S. Woodruff, an evolutionary biologist at the University of California at San Diego, one of the scientists who worked on the chimpanzee paper, which was published in 1994 in the journal *Science*. "He's been pushed upstairs to do the administration, he's going through the motions and he's not scoring hits."

But while the upstart young Turks may be notably fecund, other males who ignore the totem pole altogether nonetheless manage to insinuate themselves into female company. "Some of the males opt out of the dominance hierarchy, and they focus instead on being more affiliative with females, grooming them, staying by their side," even when the females are not fertile, said Dr. Phillip A. Morin, an author of the chimpanzee paper and now a researcher at Sequana Therapeutics, a biotechnology company in La Jolla, California. That attentiveness bears fruit, for when the well-groomed females come into estrus, the males they choose to mate with often are their faithful companions.

In her research with olive baboons, Dr. Smuts has found that a group of the animals may or may not establish a dominance hierarchy, depending on the local conditions. The baboons lucky enough to find themselves in an expanding population with abundant resources are less likely to worry about who gets what. The males, anticipating a long and productive life span, eschew haggling over rank and instead cultivate relationships with females or form amicable coalitions with other males who may help out in the future.

By contrast, when the monkey population is under stress, numbers are declining and a male can expect to die young, he is likely to take the high-risk, high-payoff strategy, devoting his energy to beating back other males and attempting, however fleetingly, to stockpile the females for himself. "Baboons come into the world with a variety of potential strategies," she said. "One or another of those strategies may be activated depending on the demographic conditions. It may be that a male has a mechanism for evaluating the probability of his own longevity." When life is ephemeral, it pays to be a general.

—NATALIE ANGIER, April 1995

Orangutan Hybrid, Bred to Save Species, Now Seen as Pollutant

IGNORE THE ABSURD NICKNAME "Junior": this scowling, hulking he-ape, his fur like flame and his belly like Buddha's, his face ostentatiously swollen with the fatty cheek flaps and throat sac of a fully mature male, is not the sort to elicit clucks and kootchie-koos from the human primates who watch him through the Plexiglas of his zoo enclosure. As the oldest male orangutan at the National Zoo (29 years old), and the father of five other orangutans here, Junior looks the perfectly feral ambassador for his species, the great red ape, a piece of the jungle caged but never conquered.

In fact, Junior, or Atjeh as he is more formally named, is an entirely zoo-made creation, a version of orangutan that almost certainly would never be found in the forests of Indonesia, the ape's native home. He and about 80 other orangutans in captivity around the country are zoo-bred crosses between two subspecies of orangutans, those originating in Borneo and those from the neighboring island of Sumatra. They are called hybrids, or "cocktail orangutans," or simply mutts, and they are history.

Many scientists lately have decided that Sumatran and Bornean orangutans are so genetically distinct they may even qualify as separate species. They are more genetically different from one another than lions are from tigers, or chimpanzees are from the more gracile bonobos, said Dr. Stephen J. O'Brien, a molecular geneticist at the National Cancer Institute who has studied orangutan DNA.

As a result of the recent molecular work, the Indonesian government and the organization that oversees zoo programs in the United States has called a halt to interbreeding Sumatran and Bornean orangutans and to allowing the current crop of hybrids to reproduce. Despite the endangered status of orangutans in the wild, the cocktails will not be considered a genetic reservoir for possibly replenishing wild populations. Atjeh and others like him have been vasectomized, tubally ligated or implanted with heavy-duty birth control treatments. The hybrids will serve out their time in zoos (which can be a long time, for orangutans live up to 60 years), but as sexual beings they are pongo non grata.

Yet while most primatologists agree this decision is best for the future of orangutans, some scientists have lately begun to attack the policy as an ape version of racism. They say that the desire to preserve the purity of the two orangutan subspecies reflects a sentimental view of nature in which humans are ever in search of the pristine, the true, the Edenic. The critics worry that hybrids sometimes are treated as second-class apes, and they point out that on occasion the animals have been removed from the orangutan displays, as though their presence there would compromise the educational mission of zoos to emphasize conservation and species integrity.

They criticize the molecular data as incomplete and misleading, and at least one geneticist said his new analysis showed the two orangutan populations in fact were closer genetically than other researchers had concluded. In fact, they attack the general tendency to turn to molecular biology for solutions to all of life's quandaries, in conservation work as much as in human medicine.

"We're experiencing a generalized fad of looking for a genetic fix on everything," said Dr. Anne E. Russon of York University in Toronto. "The data used to assess genetic differences are not extensive and we can't be certain of what the differences mean. There are some suggestions that the variation within the subspecies is the same as that between Bornean and Sumatran orangutans."

The debate about what to do with hybrid orangutans was thrashed out recently at a meeting of the American Association for Advancement of Science in Atlanta. The implications of the argument extend beyond the fate of orangutans. Scientists have argued over whether the remaining red wolves, for example, are true wolves, or whether they have been intermixed with coyotes for so long that they no longer constitute an autonomous species and therefore might not merit the investment of conservation dollars to reintroduce them back to their home range in the Southeast.

Similarly, a number of farm groups in Wyoming and Idaho have petitioned the United States Fish and Wildlife Service to remove the gray wolf from the endangered species list because some animals may be wolf-coyote hybrids.

Researchers have questioned whether it is ethical to interbreed the highly endangered California condor with the commoner Andean condor, or the rare Florida panther with its more abundant cousins in the West, even if the genetic shuffling might add vigor to an imperiled population of animals.

Dr. Terry Maple, director of Zoo Atlanta and a speaker at the orangutan meeting, complained that his zoo had recently spent a large sum to purchase what zoo officials thought was an endangered Sumatran tiger, only to learn later that it might have been, as he put it, "polluted" by Bengal or other tiger genes. "If that's the case, then we won't be allowed to breed it," he said.

But the discussion about animal breeding and genetics gets particularly heated whenever primates are involved. Some biologists feel such an affinity toward the great apes that they argue that humans and the other apes should be reclassified together under a single genus, *Homo,* to emphasize the kinship people share with chimpanzees, gorillas and, more distantly, orangutans.

In the eyes of some primatologists, the question of whether humans should work to keep the orangutan lines pure is hubristic, nearly as offensive as the idea of eugenics. Even among less extreme advocates of primate rights, the question remains whether animals that are capable of producing viable offspring necessarily should be kept separate at all costs. One scientist has jokingly suggested that perhaps the solution is to ask the orangutans, who are capable of learning rudimentary sign language. If asked, orangutans might well answer that it does not matter who they mate with, so long as they do not have to spend too much time together. Orangutans are the least convivial of all the great apes, with the only sustained socializing occurring between a mother and her young offspring.

Mating between males and females is anywhere from audacious to violent. Often the females go to great lengths to attract the attention of males, displaying their genitals, swinging around on their elongated forearms in the showiest possible fashion, or even bonking an obtuse male on the head with a branch or fruit. However, a young and inexperienced male may sometimes force himself rather brutally on a female, who screams and flails in clear indication that this is not her choice. In general, though, orangutans remain fairly placid. They are clever and manually deft beasts that can learn from humans to row canoes, open locks or even cook pancakes.

About 10,000 orangutans are thought to survive in the wild, the majority on Sumatra. To the Indonesians, the apes are the people, "orang," of the forest, "utan." The Sumatran and Bornean orangutans are not always easy to tell apart. The greatest differences occur between mature males of the subspecies. Bornean males tend to have comparatively larger and floppier cheek pads, rounder faces and darker fur, while the Sumatrans have more whiskers around the cheeks and chin and curlier, more matted hair overall, and their fur generally is a bit redder.

But Melanie R. Bond, a primate scientist at the National Zoo, said that the esthetic differences were not absolute, and that individual variations among apes sometimes complicated the picture. "The problem is that for many years we thought we could tell by physical appearance alone," she said. "A certain percentage of the time we ended up right, but a certain amount of time we were wrong. That's why there are a lot of hybrids here today. We didn't set out to deliberately breed together the two subspecies of orangutans."

With the advent of molecular tools, researchers began probing the proteins and chromosomes of the apes for clues to their biochemical differences. They found that key proteins between the two showed significant discrepancies. More striking still, researchers discovered a so-called chromosomal inversion. Part of the second chromosome in one subspecies of orangutan is flipped relative to the second chromosome in the other subspecies, and this positional difference holds for all members of one orangutan population or the other.

That sort of gross chromosomal discrepancy, said Dr. O'Brien, is larger than anything seen in the various chromosomal profiles of most of the great cats. "The Sumatran and Bornean orangutans have as many molecular differences as perfectly respectable species do," he said. "I feel we should treat them as different species."

Examining the rough outline of orangutan DNA located within the mitochondria—tiny cellular structures where the body's energy is generated—many researchers have concluded that the two orangutan populations diverged at least 20,000 and possibly hundreds of thousands of years ago, going their separate ways even before Borneo and Sumatra were divided by the South China Sea. But scientists cannot say for sure that the populations have stayed utterly reproductively isolated in all that time, particularly not since humans have been in the area and possibly traded the apes back and forth. Even today, some unknown number of wild orangutans are illegally caught, kept as pets for a while and then released into the forest, with little concern over the genetic integrity of a given orangutan population.

Still, Dr. O'Brien and others say the science indicates it would be as unethical at this point to knowingly breed a Sumatran with a Bornean orangutan as it would be to cross a snow leopard with a tiger. "A rule of thumb in conservation work is, You don't want to tamper with the genetic integrity of good populations," Dr. O'Brien said.

But Dr. C. Cam Muir, a geneticist at Simon Fraser University in Burnaby, British Columbia, presented to the Atlanta meeting data on orangutan genes that contradicts or at least complicates previous molecular studies. Dr. Muir gathered DNA samples from wild apes, collecting orangutan hair and feces and then isolating the genetic material nested within. Whereas previous studies had looked at mitochondrial DNA through a fairly blunt method called RFLP analysis, Dr. Muir decided to look at the precise chemical sequences of five orangutan genes.

"The advantage of the sequence approach is you get higher resolution," he said. Taking various statistical paths to interpreting the data, he concluded that the two subspecies sat on the same branch of the phylogenetic tree, rather than on two different branches as others had insisted. This may sound like a strange argument for a geneticist to make, he said, "but I don't think genetics is a legitimate method for distinguishing populations."

Dr. Muir's work is unlikely to have much effect on the fate of the orangutans now in the nation's zoos. Not only have nearly all the zoos agreed to abide by the call to stop breeding hybrids, but at the moment they are not breeding even pure-blooded orangutans, except in rare circumstances. The apes take up a lot of zoo space, they are expensive to keep and for the near future there is little chance that any will be bred for reintroduction into the treetops of their ancestral home.

Dr. Maple admitted that zoos were bowing under the weight of being regarded as the "last refuge" for all the world's endangered creatures. And Junior's recent vasectomy obviously will not do anything to help preserve the remaining forests of Sumatra and Borneo, where his less genetically alloyed brethren still swing in joyous solitude.

—NATALIE ANGIER, February 1995

Bonobo Society: Amicable,
Amorous and Run by Females

NATURE'S RAUCOUS BESTIARY rarely serves up good role models for human behavior, unless you happen to work on the trading floor of the New York Stock Exchange. But there is one creature that stands out from the chest-thumping masses as an example of amicability, sensitivity and, well, humaneness: a little-known ape called the bonobo, or, less accurately, the pygmy chimpanzee.

Before bonobos can be fully appreciated, however, two human prejudices must be overcome. The first is, fellows, the female bonobo is the dominant sex, though the dominance is so mild and unobnoxious that some researchers view bonobo society as a matter of "codominance," or equality between the sexes. Fancy that.

The second hurdle is human squeamishness about what in the '80s were called PDAs, or public displays of affection, in this case very graphic ones. Bonobos lubricate the gears of social harmony with sex, in all possible permutations and combinations: males with females, males with males, females with females, and even infants with adults. The sexual acts include intercourse, genital-to-genital rubbing, oral sex, mutual masturbation and even a practice that people once thought they had a patent on: French kissing.

Bonobos use sex to appease, to bond, to make up after a fight, to ease tensions, to cement alliances. Humans generally wait until after a nice meal to make love; bonobos do it beforehand, to alleviate the stress and competitiveness often seen among animals when they encounter a source of food.

Lest this all sound like a nonstop Caligulean orgy, Frans de Waal, a primatologist at Emory University in Atlanta who is the author of *Bonobo: The Forgotten Ape,* emphasizes otherwise. "Sex is there, it's pervasive, it's critical, and bonobo society would collapse without it," he said in an interview. "But it's not what people think it is. It's not driven by orgasm or seeking release. Nor is it often reproductively driven. Sex for a bonobo is casual, it's quick, and once you're used to watching it, it begins to look like any other social interaction." The book, with photographs by Frans Lanting, is published by the University of California Press.

In *Bonobo,* Dr. de Waal draws upon his own research as well as that of many other primatologists to sketch a portrait of a species much less familiar to most people than are the other great apes—the gorilla, the orangutan and the so-called common chimpanzee. The bonobo, found in the dense equatorial rain forests of Zaire, was not officially discovered until 1929, long after the other apes had been described in the scientific literature.

Even today there are only about 100 in zoos around the country, compared with the many thousands of chimpanzees in captivity. Bonobos are closely related to chimpanzees, but they have a more graceful and slender build, with smaller heads, slimmer necks, longer legs and less burly upper torsos. When standing or walking upright, bonobos have straighter backs than do the chimpanzees, and so assume a more humanlike posture.

Far more dramatic than their physical differences are their behavioral distinctions. Bonobos are much less aggressive and hot-tempered than are chimpanzees, and are not nearly as prone to physical violence. They are less obsessed with power and status than are their chimpanzee cousins, and more consumed with Eros.

As Dr. de Waal puts it in his book, "The chimpanzee resolves sexual issues with power; the bonobo resolves power issues with sex."

All of which has relevance for understanding the roots of human nature. Dr. de Waal seeks to correct the image of humanity's ancestors as invariably chimpanzeelike, driven by aggression, hierarchical machinations, hunting, warfare and male dominance. He points out that bonobos are as genetically close to humans as are chimpanzees, and that both are astonishingly similar to mankind, sharing at least 98 percent of humans' DNA. "The take-home message is, There's more flexibility in our lineage than we thought," Dr. de Waal said. "Bonobos are just as close to us as are chimpanzees, so we can't push them aside."

Indeed, humans appear to possess at least some bonobolike characteristics, particularly the extracurricular use of sex beyond that needed for reproduction, and perhaps a more robust capacity for cooperation than some die-hard social Darwinists might care to admit.

One unusual aspect of bonobo society is the ability of females to form strong alliances with other unrelated females. In most primates, the males leave their birthplaces on reaching maturity as a means of avoiding incest, and so the females that form the social core are knit together by kinship. Among bonobos, females disperse at adolescence, and have to insinuate themselves into a group of strangers. They make friends with sexual overtures, and are particularly solicitous of the resident females.

The constructed sisterhood appears to give females a slight edge over resident males, who, though they may be related to one another, do not tend to act as an organized alliance. For example, the females usually have priority when it comes to eating, and they will stick up for one another should the bigger and more muscular male try to act aggressively. Female alliances may have arisen to counter the threat of infanticide by males, which is quite common in other species, including the chimpanzee, but has never been observed among bonobos.

Dr. de Waal said that many men grow indignant when they learn of the bonobo's social structure. "After one of my talks, a famous German professor jumped up and said, 'What is wrong with these males?'" he recalled. Yet Dr. de Waal said the bonobo males might not have reason to rebel. "They seem to be in a perfectly good situation," he said. "The females have sex with them all the time, and they don't have to fight over it so much among themselves. I'm not sure they've lost anything, except for their dominance."

—NATALIE ANGIER, April 1997

In Society of Female Chimps, Subtle Signs of Vital Status

MUCH TO THE SURPRISE of primatologists, a dominant female chimpanzee turns out to be a lot like Alan Greenspan, the all-mighty czar of the Federal Reserve. He clears his throat, and stock markets tumble. She twitches an ear, and her underlings tremble. He controls the economy without so much as passing a law. She affects her group's fecundity without so much as raising a paw. Both man and ape offer proof that you don't have to speak much, you don't even have to carry a big stick, and still you can rule the world.

Scientists have discovered that female chimpanzees, long believed to have little or no interest in pulling rank on one another, in fact form subtle social hierarchies that profoundly influence the fate and fertility of every female in the group.

Working in the Gombe National Park of Tanzania with Jane Goodall, Anne Pusey and Jennifer Williams of the University of Minnesota have determined that female chimpanzees differ far more in their individual ability to bear and rear offspring than anybody had suspected, and that one of the biggest factors influencing a female's reproductive prowess is her social status. The babies of a high-ranking female are much likelier to survive to independence than are the offspring of a subordinate chimpanzee, and the daughters of a dominant mother reach sexual maturity four years earlier than those of a low-ranking female, a spectacular advantage that can transform a dominant family into a dynastic one.

But for all the discrepancies in prospects that are tied to social status, female chimpanzees appear indifferent to displays of power. They do not brandish fallen branches or kick over oil drums, as a male will to flaunt his status. They rarely fight with one another, and when they do, it is hard to tell who won. They are not big socializers, spending most of their time alone in the forest with their dependent young. Should a female encounter a male from her group, she will make the appropriate noises of subordination, giving off a little *hnn-hnn-hnn* sound called a pant-grunt, as a low-ranking male will in the presence of a dominant male. But when two females meet, they often ignore each other.

"The big surprise here is that dominance rank is so subtle as to be nonexistent, yet it has huge impact on reproductive success," said Ms. Williams, who is working on her doctoral dissertation. The findings of the study appear in the journal *Science.*

"People had emphasized the importance of rank among male chimpanzees, because it was so easy to see," said Dr. Craig Packer of the University of Minnesota, who did not contribute to the chimpanzee study but is familiar with it. "The females looked like the nebbishes of the woods," he said.

When Dr. Pusey and her colleagues began analyzing chimpanzee behavioral patterns more carefully, they found that beneath the females' apparently distracted exteriors skulked true political animals. Using data from the renowned Gombe chimpanzee field study that Dr. Goodall began in 1960, the scientists assessed dominance by looking at all pant-grunts exchanged in a group of 10 females from 1970 through 1992. They found that while pant-grunting is not an inevitable feature of female greeting, when it did occur between a given pair, it was always in the same direction: If female A pant-grunted to female B one day, A would pant-grunt to B the following year. By considering this reliable if inconspicuous vocalization, the researchers found that the females aligned themselves into a fairly stable hierarchy of low-, middle- and high-ranking apes.

The scientists then mapped out the reproductive histories of the 10 females, a difficult task for a long-lived and slow-breeding species like the chimpanzee. In their analysis, they counted as a success any offspring that survived to five, the age of weaning. The group variability in fruitfulness proved significant.

Several of the low-ranking females had many pregnancies and births, but then lost most or all their infants, either to predation, poor nutrition, illness, accident, infanticide, or causes unknown. By contrast, the highest-ranking female of the group, Fifi, is also the most successful mother. Now

38, Fifi has not lost a single one of her seven offspring. Five already are independent—including two sons in their 20s, a 16-year-old daughter who is herself a mother and a 12-year-old daughter who is pregnant—while the youngest two are on the cusp of weaning. The average age of reproductive maturity for a chimpanzee is 13, but the daughters of high-ranking females began breeding as young as 9.

The scientists do not yet know what distinguishes a high-ranking female from a subordinate, or why the offspring of dominant females fare better than those of the lowly. Dominant females are not any bigger or fleshier than their underlings, nor are they more overtly aggressive. High rankers tend to be somewhat older than their subordinates, and a few females manage to gradually gain in stature as they age, but most stay put on the social pyramid no matter how long they live.

The scientists suspect that dominant females exert their preeminence in their choice of feeding terrains. Each female has her own turf that she forages through every day, and the dominants could be monopolizing the spots where the food is particularly nutritious. In addition, because feeding ranges overlap, the researchers theorize that should two females bump into each other while foraging, the high-status female gets first dibs on any good fruits or nuts that may be around.

"One thing we're trying to look at now is whether dominant females have a more steady body size," said Ms. Williams. "It could be that they don't lose weight in poor times." Fluctuations in body weight could affect fertility, lactation and the ability to fend off disease.

The scientists propose that a female's social rank is determined fairly early in life. Some daughters stay in their natal home, and may inherit their station from their mother, as Fifi did from her regal mother, Flo. More often, a young female will emigrate from her birthplace and join another group. It is during those early months of integration that a female makes her greatest effort to seek high status and is likeliest to engage other females in fights, or at least earnest, cacophonous discussions. After a while, a female appears to learn her place and from then on will grunt or be grunted at accordingly.

The researchers caution that the sample size in their study is extremely small, and that much more needs to be done to grasp the nuances of female competition in chimpanzees. Other research on primates has shown that dominant behavior, when taken to extremes, can backfire. Studying olive baboons, for example, Dr. Packer has found that the most dominant and aggressive females can suffer from fertility problems, possibly as a result of an excess of masculinizing hormones like testosterone.

Kathleen B. Kerr, a researcher and therapist at the Georgetown Family Center in Washington, has also been analyzing the Gombe data on female chimps as part of an effort to understand mother-infant relationships. She points out that while high rank for a chimpanzee is a critical factor in fostering successful motherhood, it is not the only factor. "There's variability between female chimps of the same rank in how they behave, what their life course is, and how many offspring they successfully raise," she said. "I don't think rank is whole story."

In her observations, the chimpanzee mothers most adept at rearing young tend to be "very relaxed but conscientious," she said. They are vigilant yet calm, staying watchful for danger while encouraging their young to explore their surroundings. In comparison, the mothers with comparatively low success at raising babies to maturity "tend to be restrictive and overprotective of their offspring," she said, "which translates into the offspring being less confident and exploratory, and more restricted in their approach to the world."

It's an old story: The strong get stronger, while the weak get weaker—if they don't get eaten first.

—NATALIE ANGIER, August 1997

Meat Viewed as Staple of Chimp Diet and Mores

IF CHIMPANZEES WERE HUMAN, their thirst for blood could be called barbaric. And if human morality applied to their practice of trading food for sex, many would spend the mating season in jail.

Researchers studying chimpanzee hunting habits are gaining new insight into the lives of man's closest animal relative. Once thought of as docile vegetarians, these able hunters forage for meat with a passion and motivation not chronicled until recently.

Dr. Craig B. Stanford, an anthropologist at the University of Southern California, has documented the chimpanzee's success in the pursuit of flesh. Given other similarities between chimps and man, he suggests that early humans could have chased game millions of years before current evidence suggests. Dr. Stanford described his research, the largest study to analyze the chimpanzee's predilection for meat, in *The American Scientist* magazine.

Dr. Stanford found that chimpanzees hunt with such gusto in Gombe National Park in Tanzania that each year they lay waste to one-fifth of their territory's population of the red colobus monkey, their preferred prey. These long-tailed victims, crowned with a thatch of red hair, are plucked with abandon from the trees where they forage near the border with Zaire.

The 45-member Kasakela chimpanzee community, living in the low-lying forest of Gombe, eats one ton of meat on average each year, said Dr. Stanford, a 38-year-old associate professor at U.S.C. During one hunting binge in 1992, they killed 71 red colobus monkeys in 68 days.

This level of predation is surprising to many primatologists, but seems less so when compared with the diet of humans, the only other primate known to eat meat regularly. According to Dr. Stanford, chimpanzees can consume up to a quarter pound of meat a day when they hit their hunting stride, rivaling at times some contemporary tribes of hunter-gatherers.

"Every article ever written on human evolution up until now says, 'Isn't it fascinating that chimps eat meat, but it's trivial compared to what modern humans eat,'" Dr. Stanford said. "Well, the Pygmies of Zaire are among the lowest meat consumers on the human spectrum. And there is no question that chimpanzees are, in some months, very close to that amount or are already there."

Dr. William C. McGrew, an anthropologist at Miami University of Ohio who studies chimpanzee behavior, said, "You pick up any textbook and it would say that meat eating by chimpanzees is insignificant. This is the first time meat has been shown to be important nutritionally."

Among the first scientists to debunk the myth of chimpanzees as banana-eating vegetarians was Dr. Jane Goodall, a British primatologist, who nearly 30 years ago announced to a startled scientific community that chimpanzees were part-time carnivores. Further research showed that these natives of equatorial Africa kill fellow chimps, use tools and mourn their dead—all behaviors once thought of as uniquely human.

As early as the 1960s, Dr. Geza Teleki, an American primatologist, said after observing male chimps swap meat for sex with females that nutrition was only one of several reasons chimpanzees ate flesh.

Dr. Stanford builds on this finding, saying that male chimpanzees often hunt as a way to finance their sexual barter when traveling with sexually receptive females. And the more such receptive females are present, the more likely a group of chimpanzees will hunt.

Time after time, Dr. Stanford documented how male chimpanzees dangle a dead red colobus monkey in front of a sexually swollen female, sharing only after first mating. He said that human sexual relationships could have been just as material-based.

"When chimps arrive at a tree holding meat on the hoof, the male chimps seem to have an awareness that, 'Well, if I get meat I will maybe get more copulations because the females will come running over once I get a carcass,'" Dr. Stanford said.

Female chimpanzees are sexually promiscuous, with or without meat, copulating with more than a dozen males each day. But Dr. Stanford believes the attraction of flesh, consumption of which is shown by Dr.

McGrew to be linked to the survival of offspring, could give lower-ranking males a better chance at matings; or that it could be "the difference between getting lots of sex and getting lots and lots of sex."

The ruthless manner in which chimpanzees hunt monkeys is best illustrated by a 1992 kill at Gombe, one of the largest ever recorded. Two chimpanzee parties traveling with 33 members, including two sexually receptive females, converged underneath trees in which up to 25 colobus monkeys were noisily feeding on fruit.

The colobus monkeys, weighing nearly 20 pounds, shrieked with alarm as the group of male chimpanzees (females hunt only on occasion at Gombe) made their way up to the canopy. In the frenzy of battle, some colobus monkeys were killed in the trees by the usual bite to the head. Others fell to the ground only to be flailed against the forest floor by chimpanzees weighing nearly 100 pounds.

By the time the "gruesome bloodbath" ended one hour later, Dr. Stanford said, seven monkeys were being eaten. The most prolific hunter at Gombe has killed 42 colobus monkeys in five years.

"Chimps absolutely love meat and get extremely excited about hunting," said Dr. Richard W. Wrangham, a Harvard anthropologist who said in a 1990 study that chimpanzees prey on at least 25 different species of mammals. "They will wait for an hour under a tree for just three drops of blood to fall off a leaf."

Most hunting is seasonal at Gombe, Dr. Stanford found. It takes place during the dry summer months, a time when females are generally sexually receptive and when the food supply of fruit, leaves and nuts is scarce. During the winter, however, when sex is not usually an issue and food is more plentiful, chimpanzees can go several weeks without a morsel of monkey, baboon, small antelope or baby bush pig.

But male chimpanzees sometimes hunt at Gombe when no sexually receptive females are nearby, primatologists report. This is when chimpanzee politics comes into play.

Dr. Toshisada Nichida, a Japanese zoologist, described in 1992 a primate patronage system that would be right at home in the back rooms of Capitol Hill. A male troupe leader in the Mahale Mountains of Tanzania doled out meat portions to allies, while denying the rewards to enemies. At Gombe, Dr. Stanford observed similar politicking at mealtime.

"In the chimp society as in human society, being big doesn't get you everything," Dr. Stanford said. "It's being a politician. You have to know how to network. You have to use your political abilities to get what you want."

It is difficult to generalize about chimpanzee society. As an example, chimpanzees at Gombe capture mainly infant colobus monkeys using a pell-mell strategy, Dr. Stanford said. By contrast, Dr. Christophe Boesch, a Swiss primatologist, says male and female chimpanzees at Tai National Park in Ivory Coast prey mostly on adult colobus monkeys in coordinated attacks.

The difference in technique, Dr. Stanford says, could be accounted for by the local vegetation. The tall, densely packed trees of Tai provide more escape routes for the monkeys, and therefore require more organized hunting.

"This demonstrates that chimps have behavioral plasticity," said Dr. C. Owen Lovejoy, an anatomist and anthropologist at Kent State University in Ohio. "It's very exciting news that chimpanzees are perfectly capable of enhancing their own fitness for success."

Dr. Stanford says chimpanzees are such efficient killers, successful nearly 90 percent of the time when 10 or more males are present, that he wonders whether early man was also quite skilled.

Archaeological evidence indicates that humans hunted at least 2.5 million years ago, based on stone meat-cutting tools found at Olduvai Gorge in Tanzania. But Dr. Stanford said that from what he learned from watching chimpanzees, he believed that humans were avid hunters nearly three million years earlier than remains suggested.

"The amount of meat chimps eat suggests that early hominids, who would have presumably been more intelligent and better able to coordinate their actions and hunt together, were probably eating as much meat as chimps or more," he said.

Molecular biologists estimate that 98 percent of a chimpanzee's DNA matches that of humans. And since both humans and chimpanzees eat meat, Dr. Stanford said he thought it likely that so did their common ancestor who lived some 6 million years ago, when a branch in evolution created early man.

Scientists have been debating for some time when early man began hunting, believed by many to be a hallmark of human evolution linked to brain expansion. Some contend these human ancestors were mainly scavengers, too weak and slow on two legs to have hunted successfully, while others say Dr. Stanford's theory lacks evidence.

A team led by Dr. Timothy D. White, a paleontologist at the University of California at Berkeley, last year uncovered in Ethiopia a partial skeleton of the earliest hominid yet found, dating back 4.4 million years. Dr. White said that the hominid, known as *Australopithecus ramidus,* was probably capable of hunting but that the remains provided no concrete evidence that it ever did so.

"Stanford's work is very provocative," Dr. White said. "And it's completely plausible; but it's totally unsupported by empirical evidence. That's why we're trying to find some kind of smoking gun, though I don't know what that would look like."

Dr. Stanford said it was unlikely that evidence to support his theory of early hominid hunting would ever be uncovered. To illustrate the point, he held the bone remains of five colobus monkeys (equal to about 60 pounds) in the palm of his hand—leftovers collected after a chimp feast.

"Chimps eat hair, skin, bones—there's nothing left," Dr. Stanford said. "Early hominids probably ate everything and you wouldn't find it in the fossil record."

—VERNE G. KOPYTOFF, June 1995

7

THE NATURE OF THE MAMMAL

Mammals are named, not very gloriously, after a mere gland, the mammary gland or breast. But there is much in the name because nursing, rearing and parenting are defining mammalian activities.

Unlike crocodiles or turtles, mammals do not walk away from their progeny, leaving them to cope as best they can in a hostile world. Mammalian parents devote considerable efforts to nurturing and raising their infants.

Perhaps as an extension to these family duties, many mammalian species are social creatures, exhibiting a range of interactions from hunting in groups to the elaborate social hierarchies of the rhesus monkey.

Sociality is not something that just happens. These behaviors evolved because they offered some survival value, and mechanisms to produce them have become embodied in the genetic design of mammals.

Biologists have gained insights into some of these mechanisms, such as the hormones that underlie monogamy in prairie voles.

From play to parenting, there is a spectrum of mammalian behavior into which humans fit with perfect ease.

Why Babies Are Born Facing Backward, Helpless and Chubby

CHILDBIRTH: IT'S A MIRACLE, right? One minute you're standing there, an awkward, frightened and comically proportioned woman, face twisted with the first contractions of labor, uterus lurching to your epiglottis, mind suddenly skeptical that it's feasible to do the pelvic equivalent of expelling a Crenshaw melon through your nostril, and the next minute . . .

Oh, sorry, 23 hours later, you're sitting peacefully in bed, face now suffused with a Botticellian glow, body significantly lightened, a pointy-headed, purplish newborn at your breast. It's magic, it's sacred, it's among a woman's greatest triumphs—but, holy epidural, the author of Genesis sure said it all by socking Eve with the curse of childbirth right after making her mortal.

As anybody who has ever experienced or witnessed a baby being born will readily attest, the pain of labor and delivery deserves its legendary status, ranking as it often does somewhere above, say, a root canal and below an unanesthetized amputation. Human births are thought to be so difficult because we are pushing the envelope of evolution: The baby's large cranium is nearly too big to fit through the woman's pelvis, which must remain narrow to accommodate bipedal locomotion. We are, the assumption goes, too brainy for our poor, upstanding mothers' good. All told, human births were thought to be among the most troublesome in nature, resulting in greater rates of stillborn young and maternal deaths than seen in other mammals.

But as researchers have discovered in a series of recent studies, human birth is not the most difficult in the animal kingdom by any stretch, nor is it as unusual as biologists and anthropologists long believed. In new and still unpublished work, Dr. Melissa Stoller of the University of Chicago has carried out the first detailed analysis of the birth process in nonhuman primates, in this case, baboons and squirrel monkeys. Her findings overturn

239

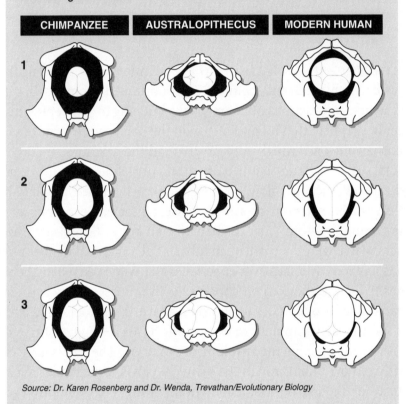

Short but Hard Trip Into the World

In these diagrams of the birth process from a midwife's point of view, the newborn's head passes through the inlet (1), midpoint (2) and outlet (3) of the birth canal; the maternal spine is at the bottom. It was once thought that among primates, only human infants had to turn sideways to navigate the canal; however, in monkeys and apes that undergo such twists, the newborns still normally emerge head first and facing the mother.

CHIMPANZEE   AUSTRALOPITHECUS   MODERN HUMAN

Source: Dr. Karen Rosenberg and Dr. Wenda, Trevathan/Evolutionary Biology

N.Y. Times News Service

the long-held assumption that most primates plop straight down through the mother's pelvis and that only human babies are forced during birth to turn their way through a harrowing corkscrew bend to take advantage of the widest possible points in the woman's birth canal. Instead, Dr. Stoller discovered that baby monkeys must also twist their way through the birth tunnel with contortionist efforts, although they rotate in different directions than humans do as they struggle toward the world.

Moreover, the ratio between the size of the infant's head and the diameter of the mother's pelvis turns out not to be as miserably incompatible for humans as it is for some other monkey species.

Nevertheless, humans do retain a few singular elements in their style of giving birth. Dr. Wenda Trevathan, an evolutionary anthropologist at New Mexico State University in Las Cruces and the author of the classic text *Human Birth: An Evolutionary Perspective,* argues that much of the pain and anxiety surrounding human birth is cognitive in nature and that it is not necessarily a bad thing.

Unlike other mammals, women know they will have a bloody difficult time in labor and delivery. They've seen it, they've heard about it and they've been warned since girlhood to expect it. That knowledge of impending misery, Dr. Trevathan argues, leads women to seek assistance during birth, a practice that is extremely unusual among mammals. Most expectant animals do the opposite, finding a quiet spot away from the group and giving birth in unobtrusive solitude, a practice that incidentally makes it quite hard for field researchers to witness births of the species they study. But humans almost universally solicit help, whether of a midwife; an elder female relative; on occasion, a male partner; and today, of course, an obstetrician.

Reporting in the journal *Evolutionary Anthropology,* Dr. Trevathan and Dr. Karen Rosenberg of the University of Delaware said that in a survey of 296 cultural groups around the world, only 24 offered instances of women sometimes giving birth unattended, and those cases almost always involved experienced mothers. First-timers, it seems, universally have someone in attendance.

Women are not seeking aid just for the companionship, Dr. Trevathan said. For mechanical reasons, women need help getting the infants out safely. Here is where humans differ dramatically from other primates. Among monkeys and apes, newborns normally emerge from the birth canal headfirst and facing their mother. Indeed, some of the twists and turns that Dr. Stoller reported seeing among baby baboons and squirrel monkeys as they passed along the canal put them into just such an advantageous presentation.

With the baby facing her, the mother primate can easily perform several critical maneuvers. She can pull the baby up out of the tight squeeze, and sometimes the infant will even start hoisting itself up once its arms are free of the canal, for many primates are born with highly developed motor skills. She can unwind the umbilical cord from the baby's neck should that be necessary, as it frequently is. And she can wipe the mucus from the baby's mouth to allow it to begin breathing.

Within minutes after birth, the newborn is clinging to her hairy chest and nuzzling for a nipple.

Dr. Trevathan explains that women cannot perform these essential tasks on their own, and for that we can blame bipedalism. The demands of standing upright have resulted in a pelvis that, for complex anatomical reasons, is best exited by the infant in a rear-facing position, rather than looking up at the mother. Because of this awkward presentation, the mother cannot guide the baby out without bending its spine backward, risking spinal cord injury and death of the baby. She cannot unwind the umbilical cord, nor can she clear her baby's mouth of mucus. For all these operations, any one of which can spell the difference between life or death for her child, she needs a helping hand. Thus was born the practice of midwifery, or its ancestral equivalent.

Dr. Trevathan dates the use of birth attendants to the onset of bipedalism, about 5 million years ago or so—considerably before human brain size began reaching its current considerable proportions. She views obstetrical intervention as perhaps the oldest medical specialty.

She also proposes that the emotions surrounding childbirth—the fear, the uncertainty, the desperate desire to not be left alone—are deepseated sensations, the spurs to ensure that women do not try to go it alone.

"I argue that the fear and anxiety we have may be deeply rooted in our psyche and that they served an important function in the past," Dr. Trevathan said. "They're natural instincts, as basic to us as love or anger, and they're not going to go away with education, or being told 'Don't worry, you can always have a C-section.' "

Human childbirth can also be viewed as one of the early examples of need-driven cooperative behavior, with one woman agreeing to help with another's birth in return for similar assistance when she is in labor. "What we see here is a prime example of a cultural solution to a biological problem," Dr. Rosenberg said. That obstetrical give-and-take can even be seen as a driving force in the evolution of intelligence, just as the needs of cooperative hunting have traditionally been.

The use of birth assistants has very likely accounted for the relatively low rate of maternal deaths and stillbirths seen historically in normal deliveries. Serious problems can arise when babies attempt to come out feet first, or breech, but the condition is equally common, and dangerous, among nonhuman primates and other mammals.

What counts as a backward birth differs from one species to another, though. Dr. Lawrence G. Barnes, curator of vertebrate paleontology at the Natural History Museum of Los Angeles County and a researcher who studies the evolution of whales, said that whale calves sometimes come out of the birth canal headfirst, rather than the proper tail-first presentation. That is fatal for the neonate because its face is exposed to the water as the rest of its body is still emerging, so it cannot get to the surface before drowning.

Danger also arises for human mothers when they give birth under severely unhygienic conditions, as they frequently did in the 18th and 19th centuries. At that point, male doctors had taken over the business of birth assistance from midwives, and they often brought to the mother's bedside infectious microbes from their nonpregnant sick and dying patients elsewhere.

Today, septic conditions and poor medical care plague pregnant women in underdeveloped nations, resulting in huge numbers of unnecessary deaths. Last month, UNICEF reported that 1 in 13 women in sub-Saharan Africa and 1 in 35 in South Asia dies of causes related to pregnancy and childbirth each year, compared with 1 in 3,200 in Europe, 1 in 3,300 in the United States and 1 in 7,300 in Canada.

The rapid growth of the human brain, or encephalization, in the last 2 million years or so has added to women's birth trauma. Yet the near mismatch between the human pelvis and infant head is not the unique and severe problem that many imagine. It is true that among the great apes—chimpanzees, orangutans and gorillas—the female pelvis is quite roomy, and birth is far easier than it is for humans. But that is not the case for small species of monkeys, which have comparatively large brains without the bulk of adulthood seen in the apes. Among squirrel monkeys, for example, the fit of the birth canal is so tight relative to the baby that, as Dr. Stoller notes in her doctoral dissertation, labor obstruction is quite common. In one colony in Alabama, 16 percent of the infants are stillborn, and 34 percent of the young born alive die soon after birth, often of birth-inflicted injuries.

In shaping birth patterns, evolution seems to push to the limits of viability when there is a difficult problem to solve. Miles Roberts, head of the zoological research department at the National Zoological Park in Washington, said the most extreme ratio in primates of large infant size to narrow maternal birth passage was among the tarsiers, a small, insect-eating monkey found in Indonesia and the Philippines.

In this case, the fetus must grow a very large brain during gestation because it will have to start foraging for itself within a month. Insects are simply too energy-poor a food source to allow the mother to nurse the infant at length, as many primates do. "The infant has to get to work quickly in the three-dimensional space of an arboreal habitat," Mr. Roberts said. "That means it needs extensive motor coordination, which means it has to start out with a big brain." Tarsier birth is so difficult and improbable, Mr. Roberts said, that once the baby is out, "you can't imagine how the mother ever gave birth to such a huge thing."

The young tarsier is said to be born at a precocial, or relatively mature, stage, with much of its brain development complete. Most animals that gestate for a long time are precocial. In the elephant, for example, pregnancy lasts 21½ months, said John Lehnhardt, assistant curator of mammals at the National Zoological Park. When the calf emerges from the womb, the mother gives it a kick, shaking the mucus from its trunk, and within minutes the animal is on its feet, ready to move. If it were helpless at birth, it would immediately fall prey to predators.

The spotted hyena, which surely deserves the Eve Memorial Medal of Honor for Withstanding Childbirth Hell, offers another striking example of a highly precocial species. In that animal, gestation lasts 120 days, extremely long for a predator, and the infant comes out comparatively large, mature and ready to kill, its jaws snapping, its teeth fully erupted. But the mother must give birth to those killers through an extraordinary organ the size and shape of a penis; the genitals are masculinized because the females are sculpted in the womb by high levels of androgens, or male hormones. First births are often deadly, as the penislike organ rips open as best it can yet nonetheless leaves many fetuses lethally trapped within the tube.

Dr. Laurence G. Frank of the University of California at Berkeley has estimated that the cost of this bizarre system is enormous: 65 to 70 percent of firstborn young and up to 18 percent of first-time mothers die. Yet the benefits of giving birth to extremely aggressive young outweigh the drawbacks. Mothers that survive have an easier time from then on, and their infants can thrive: Spotted hyenas are among the most successful African carnivores.

Mammals with large litters usually have short gestation times and give birth to quite helpless, unformed-looking creatures. Such species are said to be altricial, and they include cats, dogs and many rodents.

Humans are unusual in having a very long pregnancy time, comparable to that of the great apes, yet they give birth to infants that are far more immature, helpless and mewling than an infant chimp or orangutan. This system is evolution's solution to the big brain–small pelvis dilemma. The baby's head and brain grows as much as it can in the womb, and then it continues developing long after birth. Human babies are thus believed to be secondarily altricial: Our early hominid ancestors probably gave birth to more precocial young, but an extended period of altricial helplessness was superimposed over that to permit a burst of postutero brain growth. Newborn humans are essentially fetuses for another nine months after birth.

Given the infant's protracted period of total helplessness, human parents must diligently attend to their needs far longer than any other primate parents. Some evolutionary biologists propose that some of the last touches made to infants in the womb are essentially cosmetic, turning the baby into something so cute the parents will feel compelled to care for it.

Dr. Sarah Blaffer Hrdy of the University of California at Davis suggests that the adorable factor accounts for why the human infant puts on layers of fat right before birth, while the fetuses of great apes remain quite trim for their odyssey through the pelvis. "You have to ask why the baby doesn't wait until it's out of birth canal to plump up like that," she said. Perhaps a pinched and scrawny infant would lack the esthetic appeal needed to seduce its parents irrevocably. And once the mother has fallen in love, she forgets the pain, she forgets the hassle and she gladly accepts indentured servitude for the next 18 years.

—NATALIE ANGIER, July 1996

The Purpose of Playful Frolics: Training for Adulthood

Vertical leaping

Pirouetteing

ALONG WITH LOVE and a good joke, playfulness seems like something that should not be explained, a brilliant splash of animated joy so sheerly pleasurable to watch and engage in that it is its own justification.

Yet to scientists, the problem of how play evolved and why many young mammals, birds and even a few fish and reptiles clearly love to have fun is neither easy nor self-evident. Scientists have long known that play is widespread in the animal kingdom, and they have described their unabashed glee at the sight of a whale calf rolling and somersaulting around its mother's fluke with the viscous, goofy movements of an aquatic elephant, or a young brown bear plucking a flower with its teeth and scampering off across a meadow like a flirtatious Spanish dancer.

Only lately, however, have they appreciated just how profoundly important play must be to an animal's physical and mental growth. In reports in the journal *Animal Behaviour,* researchers have revised upward their estimates of how much energy a young animal devotes to random and apparently purposeless activity, the standard if not exactly merry definition of play. Among pronghorn fawns and Norwegian rats, for example, the young

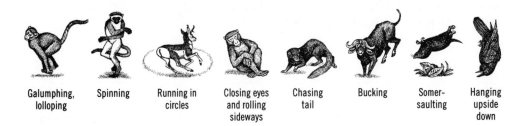

Galumphing, lolloping | Spinning | Running in circles | Closing eyes and rolling sideways | Chasing tail | Bucking | Somersaulting | Hanging upside down

Patricia J. Wynne

spend 20 percent of calories not needed simply to keep them alive on play, considerably more than traditional figures of a few percent at best.

Of greater significance, researchers recently have realized that animals absorbed in what appear to be simple displays of youthful elan—leaping, jousting, pouncing, nipping necks, chasing phantom feathers—are in fact taking enormous risks to their health and safety. While playing, young creatures expose themselves to predators, attempt potentially dangerous maneuvers through treetops or near water, and burn up fuel that might otherwise be used to get bigger faster. Scientists thus reason that evolution would never have permitted the little beasts to be so frisky were friskiness not critical to the animal's growth and performance.

"Whether it's getting stuck in the mud, or falling out of a tree, or mercilessly pursuing in an honest sense of fun a couple of weak and sickly litter mates, play is costly, particularly in terms of individual survival," said Dr. Robert M. Fagen of the University of Alaska in Juneau, one of the leaders in the field of animal play behavior.

Armed with that newfound respect for the perilousness of play, scientists lately have taken a more sophisticated approach to their research, moving beyond impressionistic observations to more rigorous studies of the physiology and psychology of play—what happens to the creature's body, brain and behavior as it revels in its escalating prowess.

They have gathered evidence that an animal plays most vigorously at precisely the time when its brain cells are frenetically forming synaptic connec-

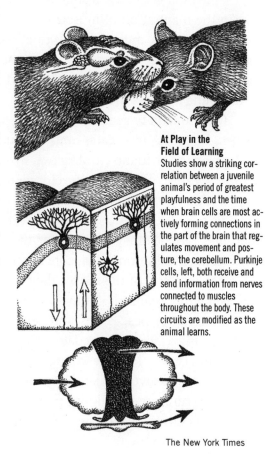

At Play in the Field of Learning
Studies show a striking correlation between a juvenile animal's period of greatest playfulness and the time when brain cells are most actively forming connections in the part of the brain that regulates movement and posture, the cerebellum. Purkinje cells, left, both receive and send information from nerves connected to muscles throughout the body. These circuits are modified as the animal learns.

The New York Times

tions, creating a dense array of neural links that can pass on electrochemical messages from one neighborhood of the brain to the next. As it turns out, the neurons sprouting especially high numbers of synapses during an animal's days of frivolity are located in the brain's cerebellum, a cauliflower-shaped region in charge of coordination, balance and muscle control. Scientists believe that the intense sensory and physical stimulation that comes with playing is critical to the growth of these cerebellar synapses, and thus to proper motor development.

And they suspect that other parts of the brain benefit from play stimulation as well, which is the likely reason why such big-brained species as primates and dolphins are outstandingly playful: In these creatures, the brain continues to flower long after birth and hence needs as much tweaking from the outside world as possible.

Some biologists have also found that the vigorous movements of play help in the maturation of muscle tissue, in ensuring the proper combination of fast-twitch muscle fibers, to allow rapid muscle contraction, and slow-twitch fibers, needed for aerobic activity. Some researchers suggest that play helps in fiber formation by sending varying types of nerve signals to the young animal's muscles.

"I've looked at age distribution of fiber development for a variety of species, for mice, rats, cats, even for giraffes," said Dr. John A. Byers of the University of Idaho in Moscow. "It turns out that the age when you see the greatest fiber tract development in these animals is just when they're at the maximal rate of play."

But researchers emphasize that while the practical needs of a growing brain and muscles may account for the genesis of play, the behavior has since assumed additional, far more subtle purposes. Through play, animals can rehearse many of the moves they will need as adults, and in different species highly ritualized pastimes have evolved to suit very different needs.

Young antelopes, lambs and other herbivores play games of mock flight, bolting away from predators that are not there, a talent clearly necessary to master before the young creatures can safely strike out on their own. Among carnivores like the great cats, wolves and hyenas, cubs pretend to capture prey: stalking, pouncing, biting, swiping at objects with claws extended. Little bats swoop around after one another in theatric arcs that are similar to the maneuvers an adult must manage to capture insects unawares. Young giant anteaters, among the most primitive mammals and none too impressively endowed with gray matter, nevertheless engage in elaborate play sequences called bluff charges. They puff out their hair like a cat, raise one front foot, and then hop menacingly to the side on the other three, roaring with all the fury of a clogged drain. They will repeat this bizarre ritual time and again at the age of about two months, presumably practicing the motions to deter predators and to keep other anteaters away from a prized anthill.

Hatchling sea turtles, while limited by their cold-bloodedness to very low-key gambols, manage to take turns holding up a front foot and vibrating it rapidly in a playmate's face, said Dr. Gordon M. Burghardt of the University of Tennessee in Knoxville, a gesture that the males will years later use during courtship.

Many species adopt play behaviors that are practice for both mating and for rearing young. In one experiment, Dr. Susan A. Brunelli and Dr. Myron Hofer of the New York State Psychiatric Institute in Manhattan examined the relationship between play and parental behavior by asking young rats to baby-sit. They took a rat of about three weeks old, the equivalent of a seven-year-old child, and put it in with a litter of newborn pups. At first, the juvenile rat tried to play with the helpless little rodent blobs, pouncing on them and trying to wrestle with them just as it would with a peer.

"It would go *kabunk! kabunk! kabunk!* pouncing onto these newborns, trying to get them going," said Dr. Brunelli.

Within several days, however, the juvenile rat, whether male or female, began softening up and acting maternally, gently retrieving a pup should it crawl away and even trying to nurse the creatures. The rodent's play behavior, it seems, is extremely plastic, and changes according to cues from the surroundings, in this case shifting from rough-and-tumble rousting to a drill for parenthood.

"By putting them in a particular context, with particular stimulus, you can get them to integrate the components of behavior that are appropriate for the context," said Dr. Brunelli.

Among highly social species, play assumes the added task of easing an animal's passage into group life, where overly selfish or hostile tendencies must be tamed if not eliminated. In rhesus macaques and squirrel monkeys, for example, young animals from three months of age on spend perhaps half their waking hours at play; and the more an animal plays, the better its chances of becoming a well-integrated member of its troupe as an adult. The play is also sexually segregated, with males loving to wrestle and females preferring games of chase over those of touch.

"Through play bouts, an animal's aggressive tendencies are socialized and brought under control," said Dr. Stephen J. Suomi, chief of the laboratory of comparative ethology at the National Institute of Child Health and Human Development in Bethesda, Maryland, who studies primate play. "The animal learns when to submit and when to pursue, and it will learn how to lose a fight gracefully."

Monkeys that fail to play much while young may not end up as complete pariahs, said Dr. Suomi, but they are less sophisticated about maintaining alliances with other monkeys and seem to be more mechanical in their overtures to potential mates. "These monkeys aren't completely incompetent, and they usually can tolerate group life," he said. "But play seems to make the difference in quality of life, between merely surviving and truly thriving."

Among some social species the adults play nearly as much as their offspring, as a ritualized way of recementing bonds between surly creatures with excitable temperaments. The collared peccary, for example, a highly aggressive creature related to the wild boar, plays frenetic games with its herd mates several times a week. Choosing a spot well hidden from predators and rubbed clean of vegetation, and responding to some sort of olfactory play signal emitted by one or more herd members, the peccaries from the eldest to the smallest will begin mock snapping, rolling over one another, locking jaws, vaulting from one side of a fellow creature to the other. And just as quickly as the communal mayhem begins, it ceases, and the peccaries settle down for a nap.

"It's herd cohesion," said Dr. Byers. "Peccaries love everybody in their herd, and hate everybody else, and this is a way of affirming that bond."

Most adult creatures are as stodgy as the rest of us and thus do not play a great deal with one another, but scientists lately have observed examples of extensive play between parents and offspring, once considered a hallmark of being human. Studying brown bears in Alaska, Dr. Fagen has discovered that cubs play at least as much with their mothers as they do with other cubs, and their favorite sport seems to be rolling around in bear hugs. Among gorillas, mothers play peekaboo with their babies, while chimpanzee mothers make silly faces.

Scientists also are trying to deconstruct the act of playing, to determine exactly which gestures an animal uses to tell a potential playmate, "Hey, it's time to play," or, should the game get very rough, "Hey, I'm only kidding."

Dr. Marc Bekoff of the University of Colorado in Boulder said that because many of the motions involved in play are similar to those an animal uses for less humorous purposes, the creature must make its playful intentions abundantly obvious.

"If an animal is going to bite another on the neck, or mount it, the animal has to communicate, 'I'm playing. I'm not trying to eat you or dominate you,'" he said. Those messages of benign intent may take on a highly stereotyped form. When a puppy wants to play, for example, it will assume the familiar play bow, crouching forward on its forelimbs and putting its hind end in the air. "It's a distinct play signal that has a small likelihood of being misinterpreted," said Dr. Bekoff.

Dr. Sergio M. Pellis of the University of Lethbridge in Alberta, Canada, has found that rats have their own version of the play signal. One rat will first run away from another, but at some point it will flip over on its back, a sign of willing vulnerability not usually seen in any other rodent venture. Dr. Pellis and his coworkers have found that if the rat is given a drug to specifically block the neural pathways in charge of commanding the rat to flip over, its talent for enlisting others to play is severely hobbled.

"With the drug, the rat will still run away from its mate, but it won't flip over," said Dr. Pellis. "And the partner really hates when that happens. If the other animal doesn't turn over on its back, the playmate will show all the signs of neurotic, dissatisfied behavior."

—NATALIE ANGIER, October 1992

Scientists Mull Role of Empathy in Man and Beast

IN THE PHYSICS OF human emotions, love may be thought of as the strong force, binding together friends, family and couples with the tight, private energy of an atom's core. But the emotion most akin to gravity, the sensation that keeps the affairs of humanity on track as surely as the Earth wheels around the Sun, is empathy: the power to recognize the plight of another and to take on that burden as though it were built to order.

After the fists of hell had punched a hole through the center of Oklahoma City, the city recovered its wits with a thousand acts of empathy and compassion. Empathy allows one to sit in a movie theater and blubber over a death that never happened to a character who never lived; it keeps charities breathing, if sometimes wheezily; and it is the reason why, if you stand in the middle of a sidewalk with a map in hand, looking bewildered, someone is bound to ask if you are lost. The human race without empathy would be like February without May, or Manhattan without Central Park.

Lured by the complexities and contradictions of empathy, its pluckiness and its fragility, its centrality to human social affairs and its frequent, sorry lapses, a growing number of scientists are seeking to understand empathy's roots and evolution. They are investigating the machinery of empathy—what happens to the body and brain when one individual connects empathically with another. They are asking whether nonhuman animals are capable of experiencing empathy, and if so, how that skill might be manifested.

As with many areas in the field of animal cognition and emotion, the study of animal empathy is rife with dispute and confusion, not the least of which is how to define the term. Some researchers argue that a version of empathy developed with the evolution of mammals, which care for their young over a protracted period and thus require a mechanism for identifying need in others—the young—and responding appropriately. These scientists define empathy as including some seemingly fraternal behaviors that have a nearly automatic feel to them. If you see a person bump a shin into a fire hydrant, for example, you very likely will wince with vicarious pain. Such knee-jerk reactions suggest to some that empathy is an evolutionarily ancient response, its neural and physiological mechanisms in place long before the advent of *Homo sapiens* or even primates.

Other researchers argue that to experience true empathy is an act of great sophistication, requiring that one be able to run a narrative through one's mind about what happened to the sufferer to bring the individual to his or her current state, and what might be done to help. To empathize is to understand beginnings, middles and possible ends, and that sort of storytelling capacity demands considerable cognitive power, researchers said. They also believe empathy demands the ability to distinguish self from others, a skill usually judged by seeing if an animal recognizes itself in a mirror. By such measures, the only species likely to rank as empathic are those we usually judge most humanlike in their behaviors: chimpanzees, bonobos (also called pygmy chimpanzees), gorillas, orangutans and, possibly, dolphins.

Empathy may be the emotion that mediates some of the altruistic behavior seen in humans and other animals, although much of that behavior may be nothing more than cold calculations of future benefits for the "altruist."

Proving empathy in animals that are not capable of declaring, "I feel your pain," is difficult, but the evidence is tantalizing nonetheless. Dr. Frans de Waal, a primatologist at Emory University in Atlanta and author of a book by Harvard University Press on animal morality titled *Good Natured: The Origins of Right and Wrong,* has gathered what he calls "stories of remarkable instances of empathy." He cites the example of a female and a male chimpanzee at the Arnhem Zoo in the Netherlands. One day the female was attempting to get at a tire hanging from a pole, because the tire held a reservoir of water and she was thirsty. However, there were other tires in the way of the water-filled one, and as much as she struggled to push them aside she could not get to her desired object.

After half an hour of failures, she walked away dejectedly, at which point the male, who had been watching her efforts the entire time, set to work. He pulled down the intervening tires one by one until he reached the rubber grail. Lifting the tire from the pole, he took it to the female, left it with her to drink from, and knuckle-walked away. "That whole thing tells you the chimp could put himself into her position and see what kind of problem she was trying to solve, a problem that he knew how to solve," said Dr. de Waal.

Beyond animal compassion, researchers are considering empathy's role as the foundation of human morality, its usefulness in encouraging co-operation among strangers or provoking guilt in wrongdoers. At the same time, researchers acknowledge that even a generally desirable state like empathy can have its ugly consequences. Dr. Martin L. Hoffman, a professor of psychology at New York University and author of *Empathy, Justice and Moral Internalization,* pointed out that one drawback of empathy is that people tend to empathize most readily with those who are similar to themselves in appearance, social circumstances, behavior and the like. "To the degree that one is very empathic toward one's own group, that may mean one is very hostile toward another group," said Dr. Hoffman. "So you get this paradox of empathy as a source of racism." Empathy encourages group identification, and groups often persist by pitting themselves against despised others.

But empathy does not always lead to cooperative bliss even within a given social group. Dr. Leslie Brothers, an associate clinical professor in psychiatry at University of California at Los Angeles, noted that humans are so primed to watch for nonverbal social signals in others, priding themselves on their empathic ability to slip easily into another person's head, that they often overinterpret body language and facial expression and impute negative meanings and motives where none exist. "We come to conclusions about the social signals we're reading, and it's easy to overemphasize our ability to accurately perceive what others mean," she said. "As a result, there's so much misunderstanding that goes on, with paranoia being the extreme example."

Moreover, it is possible to be overempathic, so easily troubled by another's troubles that one becomes paradoxically fixated on one's own discomfort and seeks to escape from the person in pain. "One study showed that highly empathic nurses tend to avoid terminally ill patients," said Dr. Nancy Eisenberg, a psychologist at Arizona State University in Tempe.

In tracing the roots of empathy, researchers have considered a range of affiliative behaviors that may be the substrate on which full-bodied empathy is built. On the primitive end of the scale is mood contagion, the passing of an emotion or behavior through a group of animals. Mood contagion causes a herd of antelopes to start fleeing, wolves or dogs to start howling in unison or people to start yawning when they see (or hear or read about) another person yawning. Contagious mood may explain why pets so often seem to their adoring owners to be empathic: A dog picks up on his mistress's sadness and begins whimpering, for example. Infants in a neonatal unit will start crying when they hear another baby crying. Nor is it the loud noise alone that is upsetting them, said Dr. Hoffman. In studies using synthesized baby cries, or white noises of the same decibel as a cry, infants did not catch the crying bug to nearly the same degree as when they heard a real infant howl.

Higher on the empathy pyramid is something akin to social smarts, said Dr. Brothers, the ability to get hints about another animal's intentions from its gestures and behaviors. In parallel with the evolution of group living, primates began developing the neural and visual equipment necessary to read one another's facial expressions and body language. As primates evolved, said Dr. Brothers, the visual system became ever more connected to the emotional centers of the brain like the amygdala, the better to link what one sees in another to what one feels. "The amygdala became more and more connected with the visual senses and less with the sense of smell, and that kept step with primates having evolved to send each other visual signals like facial expressions," she said. At the same time, the auditory pathways linked up with the brain's emotional headquarters, the better to interpret vocalizations and eventually speech.

Many types of primates, apes and monkeys alike, use facial expressions to communicate with one another. When confronted with a stranger, a rhesus monkey will open its mouth and chatter its teeth, the better to say, "I don't like you and I wish you would leave."

Going beyond a mere reading of others is a mimicking of others, a skill that Dr. Robert W. Mitchell, a psychologist and animal behaviorist at Eastern Kentucky University in Richmond, calls "kinesthetic-visual matching," or imitation. "Monkey see, monkey do" is often used as a put-down, imputing stupidity, but in fact the ability to imitate another animal's movements or expressions ranks as a mark of intelligence, and monkeys are not very good at it, while some apes often are (hence the term "to ape"). For humans, at least, engaging in facial imitation can lead to an empathic experience. That is because the muscles of the face are connected to the emotional centers of the brain. If you see somebody frown and you frown in response, you will feel more negative than you had a moment before; the opposite is true with a smile.

As it turns out, one animal that is especially adept at kinesthetic-visual matching is the dolphin, said Dr. Mitchell. If a sea lion is put in a tank of dolphins, the dolphins will start attempting to mimic the body posture of the sea lion. Dolphins swimming with human divers will coordinate the rate of their air bubbles to match those percolating from the diver.

Yet it is one thing to read and react to a variety of social cues, and another to express true empathy for a comrade in crisis. Many scientists believe that to truly empathize with another demands that one perceive oneself as an individual and relate personally to another creature's plight. To test whether an animal recognizes itself as itself, scientists give an animal a mirror and see how it reacts. Among monkeys, the response is clear and consistent: They do not recognize themselves in the mirror, instead treating the mirror image as another monkey and reacting with aggression or fear.

By comparison, the great apes like chimpanzees and orangutans show intriguing indications of self-recognition when given a mirror. They preen, they open their mouths and inspect their teeth; they use the mirror to look at their backsides, they play with their genitals and delight in their exhibitionism.

To further test for self-recognition, scientists have anesthetized chimpanzees and other apes and put a dot of paint somewhere on each animal's face. When again confronted with the mirror, the primates often try to wipe the spot off, recognizing it as a blemish they did not have a little while earlier.

Dr. de Waal said his observations indicate concordance between a species' capacity for self-recognition in a mirror and its likelihood of displaying compassionate, empathic behavior toward its fellows. Monkeys do not recognize themselves in a mirror, and they would never put an arm around the shoulder of a friend hurt in a fight, he said. Chimpanzees have been shown to do both.

He and others emphasized, however, that the meaning and validity of mirror self-recognition tests are sharply disputed among animal behaviorists. "Not all members of a particular species will recognize themselves in a mirror," said Dr. Mitchell. "Sometimes an animal will respond, but then months later it won't. The whole field is a mess right now, with everyone claiming their experiment is better than the others."

Dolphins, for example, show some sense of mirror self-recognition, but the question remains open whether they empathize with one another. They are a highly social species, and must reach cooperative decisions about when to fish, when to sleep and when to roam, and so empathy could help cement the group bonds that ease their communal negotiations.

—NATALIE ANGIER, May 1995

Animals That Are Peerless Athletes

TO THE RECREATIONAL JOGGERS of the world, champion runners like Carl Lewis and Olga Markova are the anointed children of Zeus, cardiovascular demigods who seem to ignite the air molecules around them as they race.

Yet even the most elite human athlete is thoroughly pathetic compared with nature's other aerobic masters. Take the pronghorn antelope of Wyoming, for example, a goat-size ungulate that may rank as the greatest athlete alive. Its maximum speed of 60 miles an hour is second only to the cheetah's top pace of 70 miles an hour. But while the cheetah can sprint for just a few seconds before practically collapsing, the pronghorn can maintain its freeway stride for an hour.

When migrating between California and South America, a hummingbird will fly for 24 hours at a stretch, all the while beating its wings thousands of times each minute. In so doing, the tiny creature burns up such huge volumes of oxygen so efficiently that it would take a human athlete working for a week at peak capacity without a moment's break to match the hummingbird's aerobic performance.

The world's smallest mammal, the Etruscan shrew, is almost as light as a dime, and yet it can run more than half as fast as the fastest human. Were it scaled up to human dimensions, it would dash through a mile in about 40 seconds.

Such are the conclusions that physiologists reach when they compare the aerobic capabilities of extremely athletic mammals with those of people. Naturalists have long been impressed with the swoops, soars, dives and dashes displayed by the creatures they study, but only lately have researchers begun to understand the physiological basis of the animals' feats.

They have learned how the hearts, lungs and circulatory systems of aerobic stars like the pronghorn and the blue fox differ from those of such rutabagas of the animal kingdom as the 100-pound capybara, a four-foot-

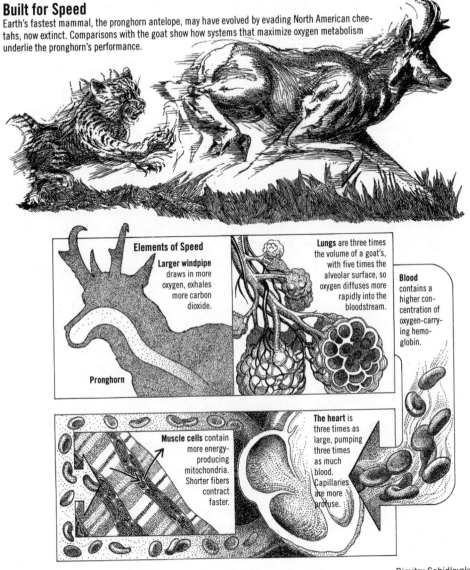

Built for Speed

Earth's fastest mammal, the pronghorn antelope, may have evolved by evading North American chee-tahs, now extinct. Comparisons with the goat show how systems that maximize oxygen metabolism underlie the pronghorn's performance.

Elements of Speed

Larger windpipe draws in more oxygen, exhales more carbon dioxide.

Pronghorn

Lungs are three times the volume of a goat's, with five times the alveolar surface, so oxygen diffuses more rapidly into the bloodstream.

Blood contains a higher concentration of oxygen-carrying hemoglobin.

Muscle cells contain more energy-producing mitochondria. Shorter fibers contract faster.

The heart is three times as large, pumping three times as much blood. Capillaries are more profuse.

Dimitry Schidlovsky

long South American rodent that Dr. James H. Jones, a physiologist at the University of California at Davis, describes as a "big ball of flab."

Some of the recent results in the field of comparative physiology were presented at a meeting of the American Lung Association in San Francisco. Speaking to an audience of clinicians concerned largely with human respiratory ailments, the physiologists emphasized how much could be learned about human anatomy and performance by considering the cardiovascular prowess of other species.

Many of the latest studies focus on oxygen transport, the journey of a breath of air as it passes down the trachea and into the lungs, through the capillary walls of the lungs and into the blood, and from the blood into the muscles, where the oxygen is transformed into energy. Animal performance depends on many factors beyond oxygen use, including muscle mass, skeletal strength and limb length, but it is the relative efficiency of oxygen metabolism that accounts for the streak of a stallion or the sluggishness of a porcupine.

"If you're trying to understand how the oxygen transport system really works, and you restrict your studies to humans, you'll see maybe a difference of only five or six percent between a sedentary person and an athlete," said Dr. Jones, a leading figure in the field. "But when you start comparing systems between animals, you come up against differences of fivefold or tenfold, and those are much easier to measure and to understand."

Comparative studies have also revealed where the limits to human athletic achievement may lie. Scientists have found, for example, that the lungs may be the biggest bottleneck to improved performance in sports. The heart can grow larger with training and thus can pump more blood around the body. The greater the volume of blood, the more oxygen it can pick up from the lungs and take to the muscles where it is needed. In addition, a rigorous training program will induce the abundant sprouting of new capillaries to more efficiently deliver oxygen-rich blood to the muscles. But the lungs themselves do not grow or improve with exercise, and eventually their fixed structure may be what holds back the swiftest possible transit of oxygen from the airways and into the bloodstream.

"The lungs of an elite athlete and the lungs of a couch potato are the same," said Dr. Peter D. Wagner, a physiologist at the University of California at San Diego. "The cardiac output of the athlete is higher, but the lungs aren't matched to the higher flow rate.

"In this sense, and in several others, champion athletes are like racehorses. As it flies fly around the track, its snout thrust forward, its eyes practically popping from its skull, the racehorse, too, is limited by the unchangeable size of its lungs."

Yet as relentlessly as elite human athletes may push the envelope of their pulmonary power, they do not show signs like those seen in thoroughbred horses that they have gone over the edge. It turns out that nearly all thoroughbreds bleed into their lungs after a race, a clear sign, physiologists said, of the dangers of breeding animals for one purpose alone. The bleeding results from the great volumes of blood coursing through the lungs, putting unbearable pressure on the lung capillaries and causing them to burst. Some horses are so fragile, they even begin to bleed through the nose during an easy trot.

"Racehorses have been selected so strongly for a single characteristic, speed, that the rest of the animal hasn't had the opportunity to evolve ways to handle it," said Dr. Stan L. Lindstedt, a physiologist at Northern Arizona University in Flagstaff who has studied pronghorn performance. "Pronghorns are much more stable. They'll flat out beat horses, but there's no indication of bleeding into the lungs. They have had millennia to work out their problems."

But horses, like many other mammals with great aerobic capacity, do have a skill that humans lack. When horses begin running, their spleens contract and spew out large volumes of red blood cells, the cells that carry oxygen. "It's a huge effect," said Dr. Wagner. "Within seventy-five seconds, the number of circulating red blood cells doubles." The only way humans can increase their red blood cell concentration is to inject more cells, a trick called blood doping, in which a person injects himself with his own or another's blood right before a race. The practice can indeed improve performance, but it is illegal in most sporting events.

In conducting their comparative experiments, the researchers have inveigled a wide spectrum of animals into galloping themselves windless on treadmills and sticking their snouts in masks, where expired air can be captured for analysis to see how much oxygen is metabolized each second. They have analyzed blood gases and taken muscle biopsies to count the number of capillaries that feed into the tissue, bearing the red blood cells rich with hemoglobin, which encages the needed oxygen.

The different species have been more or less amenable to study depending on their natural inclination for exercise. Dr. Lindstedt said the pronghorns have happily participated in the workouts. "They liked it so much that you'd open the door to the lab in the morning and they'd run right in and jump on the treadmill," he said.

Not so the goats or the steer. "They'll only run for you if you give them a ton of grain," said Dr. Jones. "They figure, 'Okay, okay, if this is what I have to do to get my treats.'"

From comparative experiments, physiologists have learned that the single biggest factor determining how much oxygen a mammal uses is its size: The smaller the animal, the more oxygen it takes in for every ounce of its flesh. "A gram of shrew muscle metabolizes fifty times as much oxygen as a gram of whale muscle," said Dr. David E. Leith, a physiologist at the Veterinary Medical College at Kansas State University in Manhattan. Why smaller animals must rev up at such a high metabolic rate is one of the great mysteries of physiology. "People used to think it was because smaller animals have more surface area relative to their weight than do bigger animals, and so they needed a higher metabolism to keep their body temperature up," said Dr. Kim E. Longworth, a research fellow at the University of California at Davis. "But the relationship between the two measurements hasn't held up, so this is still a big question in the field."

Whatever the reason for the difference in metabolic rates, scientists have worked out in general how small mammals manage to take in and use so much oxygen. A tiny animal like the shrew breathes about 50 times as often as an elephant. Its windpipes, accordingly, are quite large relative to its body size, to accommodate a high influx of air. The shrew's lungs have greater internal surface area relative to its size, and thus they can trap more oxygen. To remove the oxygen from the lungs as swiftly as possible, the shrew's heart beats 1,200 times a minute, compared with the human heart rate of about 75 beats per minute. The pumped blood is exceptionally rich in hemoglobin, the molecular carrier that shuttles the oxygen through the body. The animal's muscles are dense with capillaries to take in oxygen and expel carbon dioxide waste.

Of great importance, the muscle cells themselves are designed to put the delivered oxygen to immediate use: Up to 50 percent of their volume is taken up with mitochondria, the little powerhouse structures where oxygen, along with sugar, is converted into energy. By contrast, the muscle cells of a cow, an animal whose most vigorous activity is likely to be swatting flies with its tail, are built of only about 2 or 3 percent mitochondria. The concentration of mitochondria in human muscle cells falls somewhere between the two figures, and it turns out that this density is yet another factor that can be improved with physical activity.

Apart from the general correlation between small size and high oxygen metabolism, animals of the same dimensions often possess very different aerobic abilities. A gopher and a blue fox, for example, both weigh about 10 pounds, but the fox is far and away more efficient at burning oxygen. Examining why, Dr. Longworth has found that the key is the fox's extremely elevated heart rate. She has been able to show that in the fox, the blood is driven so quickly through the lungs that the pressure of oxygen in the lung spaces is always greater than it is in the bloodstream, a difference that ends up easing the diffusion of oxygen from the lung chambers into the blood vessels.

Comparing goats with the aerobic miracles, the pronghorns, Dr. Lindstedt found that the antelopes have no structures the goats lack, but merely have more of everything on the oxygen transport pathway. The pronghorns' lungs have five times the surface area, and thus five times the opportunity to pick up oxygen. Their hearts are three times as big and thus can pump three times the volume of blood through the body. Their tracheas are much wider, their muscle mass is greater, their concentrations of hemoglobin in the blood and mitochondria in the muscle cells are much denser. Dr. Lindstedt proposes that the pronghorns evolved the talent for speed as a way to evade now extinct ancestral cheetahs that once lived on the continent, and evolved their endurance to escape wolves, predators that hunt over long distances.

But the pronghorns pay a price for their extraordinary design. "They have virtually no body fat," said Dr. Lindstedt. "In a bad Wyoming winter, many of them die of starvation."

—NATALIE ANGIER, June 1993

Mother's Milk Found to Be Potent Cocktail of Hormones

AS SCIENTISTS LATELY HAVE struggled to learn exactly what human milk is made of, the list of ingredients has gotten so long that breast-feeding infants should be grateful their packages come without a food label.

Beyond the proteins, minerals, vitamins, fats and sugars needed for nourishment, there are antibodies in milk to help fend off infection during the early months, when the baby's own immune system is still too weak to work; growth factors thought to help in tissue development and maturation; and an abundance of hormones, neuropeptides and natural opioids that may subtly shape the newborn's brain and behavior.

Now researchers have found proof for what they have long suspected: Not only does the breast extract potent hormones from the mother's blood and concentrate them in the milk, as researchers have shown often happens; it also generates some of these hormones itself, to ensure that a rich yet precisely calibrated supply of the compounds will end up in the infant's food.

Reporting in *The Proceedings of the National Academy of Sciences,* Dr. Yitzhak Koch and his colleagues at the Weizmann Institute of Science in Rehovot, Israel, have found that a gene in charge of producing an important brain hormone, gonadotropin-releasing hormone, is switched on in the mammary glands of nursing rats, but not in the breast tissue of virgin rats. The discovery is the first detection of a neural hormone being synthesized in the breast gland proper, rather than starting out in the mother's brain or some other part of the body and ending up in the milk.

Although the experiments were done in rodents, the researchers are almost sure the results apply to humans as well. Scientists already knew that the hormone, abbreviated GnRH, exists in human milk in concentrations far exceeding the levels seen in the mother's blood, a discrepancy suggesting that a nursing woman's breast tissue generates the hormone on its own.

What exactly the hormone does for a suckling infant remains unclear, but researchers propose that it influences the development of the newborn's sex organs, forestalling their maturation until the offspring is ready for reproduction. It could also assist in the wiring of the brain regions in command of sexual behavior.

The new discovery highlights scientists' growing appreciation that the breast is not a passive udder designed simply to dispense calories to a baby, but an active gland that directs the course of the newborn's great unfolding. The breast "is a unique gland, an underestimated gland," Dr. Koch said. "Its activity is much more complex than people had thought."

Dr. Koch proposes that the breast be thought of as the external counterpart of the placenta, picking up where the large, liverlike structure left off the task of ushering the infant toward physical and neurological completion. He points out that the placenta is already known to synthesize GnRH and deliver it to the fetus, just as the breast has now been shown to do.

"They may be complementary organs," he said. "The placenta is responsible for regulating the growth and differentiation of an embryo. But after birth, not all the organs of the infant are fully developed. The brain is still growing." So, he said, "the breast could be doing a similar job" to that of the placenta.

Researchers suspect that many neural and other hormones besides GnRH will prove to be manufactured right in the breast tissue. But even in cases where the breast does not make a particular substance, and instead takes the compound from the mother's blood and concentrates it in milk, the factor in question could be integral to the infant's growth, which is why the breast saw fit to condense it in the first place.

Analyzing breast milk, researchers have found the hormone melatonin, which is thought to help the body keep time and may enable the infant to know when it should eat. They have found oxytocin, a hormone associated with affiliative impulses, which may help initiate the onset of a loving bond between mother and infant.

They have found all the various thyroid hormones, and they have argued back and forth over whether breast-feeding can alleviate and even prevent the symptoms of congenital hypothyroidism, a condition that can result in severe mental retardation. They have seen bradykinin, a small hormone involved in the sensation of pain, and they have seen endorphins, the body's natural painkillers. They have found insulinlike growth factor, epidermal growth factor and nerve growth factor, which may work either independently or synergistically in encouraging the development of the brain, liver, intestines, pancreas and other organs of the body. And this is the short list.

"Human milk is an incredibly complicated substance," said Dr. Martha Neuringer, a research associate professor of clinical nutrition at Oregon Health Sciences University in Portland. "It contains proteins we haven't even identified yet, much less know the function of."

Dr. Stephen Frawley, a professor of cell biology and anatomy at the Medical University of South Carolina in Charleston, said, "It's a cocktail of potent hormones and growth factors, most of which we're just beginning to understand."

In his own research, Dr. Frawley and his colleagues recently have discovered a new milk factor that they have named mammotrope differentiating peptide. Through test-tube experiments, they have shown that the peptide fosters the maturation of cells of the pituitary, a critical gland at the base of the brain that supplies many of the body's hormones. But the biologists have no idea how the peptide works or much else about it beyond the fact that it is small.

The complexities of breast milk are greater still. Some hormones appear only in the colostrum, the yellowish fluid that the breast secretes before the onset of lactation; others show up only later, in the milk; while still others rise and fall throughout the weeks or months of breast-feeding.

This ebb and flow may influence the timing of certain developmental events. For example, Dr. Otakar Koldovsky, a professor of pediatrics and physiology at the University of Arizona in Tucson, said that rodent studies showed that the growth protein called epidermal growth factor helped coordinate the sequence of developmental landmarks, ensuring that a pup's eyes opened at five days after birth, and the vagina of a female pup opened about 10 days after that. In rats fed on formula containing either no epidermal growth factor, or excessive amounts of it, the opening of the eyes and genitals is delayed or accelerated.

The latest studies of peptides and other hormones in milk offer yet another reason why, whenever possible, mothers should breast-feed their babies; scientists warned they were a long way from being able to chemically paraphrase in an infant formula the intricacy of human milk.

Even though most formulas are made from cow's milk, that milk is not necessarily a match for the breast, although cows do make many if not most of the hormones and growth factors that human mothers produce. For one thing, some of the hormones are species-specific: that is, their conformation and possible function vary from one creature to another, and what works for a calf may not always work for a human baby. For another, the process of pasteurizing milk will break down or deactivate many of the peptides and hormones.

Researchers are quick to note that many people have fared just fine without having been breast-fed. The bulk of the baby boom generation, for example, was raised on the bottle, and there are certainly enough of these people around to serve as evidence against any apocalyptic predictions.

In addition, many women cannot breast-feed because of illness or other physical conditions, and adopted children are formula-fed, and physicians are quick to reassure these mothers that formulas today are better than ever and that their babies very likely will thrive.

But some experts believe that the drawbacks of bottle-feeding, though subtle, may nonetheless exist, and that researchers will uncover the deficits once they begin asking the right questions. They propose that it may take years before health problems from a lack of breast milk show up as, say, a heightened risk for cancer, diabetes, or diseases of the colon.

"This is still at a fantasy level, but it could be that the study of neonatal nutrition will end up being most useful to the field of geriatrics," Dr. Koldovsky said. "It's possible that neonatal nutrition, and the hormones and growth factors you get at that point, will affect your well-being as a senior citizen."

Already there are clues that the impact of gonadotropin-releasing hormone in milk may resonate well into adulthood. Studying the effects of GnRH on the physiology of newborn rats, Dr. Sergio R. Ojeda, head of the neuroscience division at the Oregon Regional Primate Research Center in Beaverton, and his coworkers have learned that the hormone suppresses the premature development of the reproductive organs of females. Once ingested, the hormone fills in little docking sites studding a young rat's ovaries and keeps them from responding to competing signals in the body that might otherwise urge rapid maturation.

"Through breast milk, the development of the ovaries is kept in check, to keep them from becoming activated too early in life," Dr. Ojeda said. "It's an extra mechanism to insure the system works well."

Dr. Ojeda cannot say whether GnRH in human milk plays a similar role in baby girls. But he pointed out that the need for such hormonal subduing could be even greater in humans than it is in rodents. It turns out that estrogen levels in the placenta soar shortly before birth, which means that an infant girl is exposed to an estrogen spike capable of influencing her tiny sex organs.

Studies of stillborn infants have revealed that the estrogen does indeed hyperstimulate the ovaries, sometimes causing the egg follicles to enlarge to almost cystlike proportions. (The so-called witch's milk that oozes from the nipples of many newborn girls, and some boys, is another reaction to the prebirth estrogen rise.) GnRH in mother's milk could be the hormonal signal that reverses that ovarian agitation, Dr. Ojeda proposed.

If so, then the lack of breast milk theoretically could result in the premature development of the ovaries, a state that can be dangerous. Physicians know that girls who mature sexually at a young age are at an increased risk of infertility problems and breast and uterine cancer later on. In developed nations, the age of first menstruation has dropped significantly over the last 75 years or so, a phenomenon that many experts have ascribed to the West's high-calorie, high-nutrient diet. The new studies, however, suggest that bottle-feeding might also contribute to the early onset of menstruation.

But Dr. Ojeda warned that such ideas remain speculative. He said he knew of no studies comparing the relative breast cancer rates in women who were breast-fed with those who were raised on cow's milk or formula. Equally unknown is how GnRH in milk affects the growth and health of boys.

Despite the many puzzles that still surround breast milk, its power is so evident that in Sweden, at least, "it's considered unethical to feed infants anything but human milk," said Dr. Neuringer. After the birth of a child, a woman is given plenty of time off from work to nurse an infant. For those who cannot nurse, there are banks of human milk, just as there are blood banks. And the comparison is apt, for both are rivers of life whose depths scientists have yet to fathom.

—NATALIE ANGIER, May 1994

What Makes a Parent Put Up with It All?

PARENTHOOD MAY BE THE most natural task in the world, but considered objectively the job description matches that of, say, a serf. There is building a nest, struggling through labor, suckling the newborn or fetching it food, cleaning up its messes, beating back predators and fussing over its every mewl and whinny and whine. What mad potion would inspire any creature to take on such a job, and to do so with a zest that looks like . . . pleasure?

After long groping about in the dark, scientists at last are gaining clues to the biochemical basis of parental behavior. They are learning precisely what hormonal signals impel males and females to pair up into cooperative units and assume the demands of rearing and protecting young.

The work has focused largely on rodents, which display a reasonably rigid and predictable set of behaviors that can be manipulated and understood. But studies of higher animals like humans suggest that the same hormones that shape the dynamics of rodent family life may also influence human social behavior, including the responsiveness of a mother to her baby, the affectionate bond between male and female and the capacity of a child to connect with the outside world and form friendships.

The two hormones that appear to be essential to family and other social relationships among mammals are oxytocin and vasopressin, small and structurally similar proteins produced in the brain that divide in their impact along roughly, though not exclusively, sex-specific lines, with oxytocin influencing female behavior and vasopressin stimulating monogamous and paternal behavior in males.

The hormones have long been associated with tasks other than controlling behavior. Oxytocin is familiar to doctors as the hormone that spurs birth contractions and milk production in women, while vasopressin has been studied for its role in contributing to elements of the body's fight-or-flight response to stress, like raising blood pressure. But lately scientists have realized that the power of the hormones extends far beyond physiology.

"Nature is conservative, and this is a beautiful example of that," said Dr. Thomas Insel, a neuroscientist at the National Institute of Mental Health in Bethesda, Maryland, who has studied the hormones for years. "The same peptides that are important for things like uterine contractions and feeding an infant are also important for monogamous social bonds and parental behavior."

In a particularly clear and dramatic demonstration of the impact of vasopressin on behavior, Dr. Insel and his colleagues reported in the journal *Nature* on their work with prairie voles, small, fuzzy orange-brown rodents native to the Midwest. The voles are famed for their monogamous and egalitarian ways, with males and females teaming up for life and contributing jointly to pup-rearing duties. The scientists showed that vasopressin is the ingredient responsible for transforming a naive young male into an affectionate and aggressively protective partner and father.

The researchers first determined that the male's change in behavior begins with the act of intercourse. Immediately after a male has mated with a female, he shows distinct signs of preferring her to other females, cuddling with her and attacking strange voles of either sex that approach his turf.

"Aggression is one way of expressing attachment," said Dr. James T. Winslow, an author of the *Nature* paper who is now with Hoechst-Roussel, a pharmaceutical company in Somerville, New Jersey. "Parental behavior and partner preference can emerge as displays of aggression toward strangers."

The researchers demonstrated that vasopressin is responsible for the change in attitude by injecting the voles with a drug to block the hormone. Voles given the treatment maintained their sexual randiness, mating with an available female, but exhibited no particular fondness for her afterward. Nor did the males become aggressive toward strangers.

What is more, when male voles were given vasopressin treatments without having mated, they behaved toward a nearby female as though their relationship had been consummated, preferring her over others and assaulting intruders.

Significantly, vasopressin manipulations of either type had no discernible impact on the behavior of another species of vole, the montane vole, which is not monogamous. Male montane voles lack the brain circuitry necessary to respond to the calls for good paternal behavior.

In other studies of prairie voles, Dr. Geert De Vries and Dr. Zuoxin Wange, neuroscientists at the University of Massachusetts at Amherst, have demonstrated that when they injected vasopressin blockers into regions of the brain rich with vasopressin fibers, they could sharply reduce the parental behavior of father voles, inhibiting their tendency to huddle over pups, groom them and carry them safely into corners.

But vasopressin's impact on behavior is not limited to nuclear family life. Researchers in Bordeaux, France, have shown that in male rats—creatures that show little taste for pair-bonding or caring for young—vasopressin stimulates social memory, allowing males to recognize one another. Rats given vasopressin blockers fail to recognize old acquaintances and begin each encounter with a full-body sniff, a sizing up normally reserved for newcomers.

The role of vasopressin in human behavior remains in the realm of speculation, but Dr. Insel suggests that some psychiatric disorders, like autism and a type of schizophrenia, may be the result of depressed vasopressin production. Preliminary studies show that autistic children have abnormally low levels of vasopressin in their bloodstreams, but whether such measurements reflect the activity of the hormone in the brain is unknown.

Nevertheless, Dr. Insel suggests that if some way could be developed to get vasopressin across the protective barrier shielding the brain from proteins in the blood, a vasopressin mimic could help autistic patients develop the social attachments they seem so assiduously to shun. Whether vasopressin could ever be used to foster a generation of sensitive, attentive fathers, however, is another and more far-fetched notion altogether.

"In humans and other primates, a single peptide is not the entire story for social-bond formation," said Dr. Insel.

What vasopressin does for males, oxytocin seems to accomplish for females, encouraging the growth of social bonds. But Dr. C. Sue Carter, a neuroscientist at the University of Maryland in College Park and an author of the *Nature* report, points out that while vasopressin has a complex impact on an animal and elicits defensive aggression, oxytocin's effect is more frankly benign. "All the literature so far suggests that it's associated with positive social behaviors," she said. As discussed in a comprehensive book on the subject, *Oxytocin in Maternal, Sexual and Social Behaviors,* published by the New York Academy of Sciences in 1992, the hormone arouses an urge to cuddle, a zeal for mating and a willingness to care for the young.

In studies of rats and mice, neuroscientists have learned that the release of oxytocin into the brain during lactation causes the connective tissue around nerve cells to retract, encouraging the growth of new synaptic connections between one neuron and the next. "It's remarkable that brain structure can be changed at the time of the release of the hormone," said Dr. Carter. "This points out that the brain is far more plastic than most people, even scientists, realize."

In studies of women that she admits are inflammatory in their sociocultural implications, Dr. Kerstin Uvnas-Moberg, a professor of pharmacology at the Karolinska Institute in Stockholm, has found that women's scores on personality scales measuring traits like anxiety and aggression change significantly during and just after pregnancy. "They are much calmer, and more sensitive to the feelings of other people and to nonverbal communication," said Dr. Uvnas-Moberg. "They get higher points in something called social desirability, a willingness to please others."

Measuring women's blood levels of oxytocin before and during pregnancy, she and her colleagues discovered that the sharper the rise in the hormone concentration with pregnancy, the higher the woman's score on the social sensitivity scale. The levels of oxytocin remain elevated during breast-feeding, said Dr. Uvnas-Moberg, and so, too, do the women's behavioral changes as gauged by the personality tests.

The researchers do not yet know, however, if the association between the two is causative or simply coincidental. Nor do they know if the putative hormonal influence on temperament lasts beyond the months of breast-feeding. "We're only saying the changes occur when the women have children," said Dr. Uvnas-Moberg. "We don't know if they stay when the children get older. We don't dare say anything about that yet."

—NATALIE ANGIER, November 1993

Illuminating How Bodies Are Built for Sociability

LET US NOW PRAISE GENTLENESS, for as much as people must compete for status or global markets, they need sociability, affection, love. These are not options in life, or sentimental trimmings; they are part of the species survival kit.

Children who are not held or given love when young may grow up into disturbed, scared and sometimes dangerous people. Adults who isolate themselves from the world, refusing to so much as own a pet, are likelier to die at a comparatively young age than those who cultivate companionship. This unshakable dependence on others is not confined to humans but extends to any creature designed for group living. Carnivores need meat, migratory animals need motion, and social animals must socialize.

The importance among a wide variety of species of comity and friendship, grooming sessions and peacemaking gestures, and what one researcher wryly dubbed nature's original "family values," received its celebratory due at an unusual conference held at Georgetown University in Washington, D.C. The meeting, called "The Integrative Neurobiology of Affiliation," was organized by the New York Academy of Sciences to address a subject long neglected and even scorned in scientific circles: the biology of benevolence.

In the thematically broad embrace of the conference, scientists discussed the rituals of reconciliation and solace that chimpanzees and other nonhuman primates engage in after a nasty fight that threatens social ties: gestures like holding out a hand to shake and make up, or hugging and grooming, or mouth-to-mouth kissing. The researchers considered the neural and hormonal differences between the rodent species that form inseparable pairs and those that prefer to go it alone. They explored instances

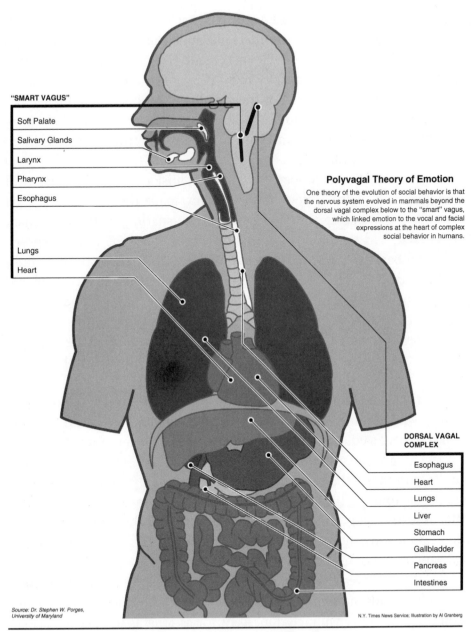

"SMART VAGUS"

Soft Palate

Salivary Glands

Larynx

Pharynx

Esophagus

Lungs

Heart

Polyvagal Theory of Emotion

One theory of the evolution of social behavior is that the nervous system evolved in mammals beyond the dorsal vagal complex below to the "smart" vagus, which linked emotion to the vocal and facial expressions at the heart of complex social behavior in humans.

DORSAL VAGAL COMPLEX

Esophagus

Heart

Lungs

Liver

Stomach

Gallbladder

Pancreas

Intestines

Source: Dr. Stephen W. Porges, University of Maryland

N.Y. Times News Service; Illustration by Al Granberg

Al Granberg

of humans who are unable to love or connect with others, the sorrowful outcome of neuropsychiatric disorders like autism and schizophrenia.

Throughout the meeting, scientists made clear what may at first seem counterintuitive, that the capacity to be pleasant toward a fellow creature is in a sense hard work. It is not the default mode. Instead, affiliative behavior requires a hormonal and neural substrate, an activation of circuitry every bit as intricate as the mechanisms controlling the body's ability to fight an opponent or flee from danger.

Dr. Kerstin Uvnas-Moberg, of the Karolinska Institute's division of physiology and pharmacology in Stockholm, made the point graphically by displaying opposing slides, one of a fierce, snarling battle-ready man, fists cocked, and the other of a nursing Virgin Mary, she of the exposed breast and benignant mien. The warrior's so-called stress circuitry is indicated and labeled. The levels of fight-or-flight hormones like cortisol and epinephrine are surging, his heart rate has accelerated, his blood pressure and blood sugar are soaring, and any gastrointestinal activity that could divert energy from his muscles has ceased. All in all, he is in a state of physiological catabolism, a mobilization and breaking down of the body's energy stores for the business of attacking an enemy.

Of the calm Madonna circuitry—the physical condition that defines a woman who is nurturing her baby—comparatively less is known, Dr. Uvnas-Moberg said, but researchers are beginning to flesh out the details. In a lactating woman, anabolism replaces catabolism: The emphasis is on building up rather than tearing apart. Insulin levels mount, the better to pull sugar from the blood and store it in cells; so, too, do the concentrations of gastric acids and hormones like gastrin and cholecystokinin, all of which aid in efficient digestion and the transfer of energy from food to the body and to breast milk.

Within minutes after beginning a bout of nursing, the mother's cortisol levels subside and her blood pressure drops, fostering a sense of relaxation that keeps her willingly quiescent for as long as it takes to sate her child; at the same time, the blood vessels of her chest dilate, which turns her into a living space heater to warm the suckling infant.

If the fight-or-flight response is seen as a strengthening of the distinction between self and the other—a tightening of the body's response mechanisms, like springs compressed into a box—then the affiliative, nurturing circuitry suggests an opening up, an expansion of self toward others, and a trading of anxiety for at least a momentary state of quiet joy.

Orchestrating this broad suite of maternal responses, Dr. Uvnas-Moberg said, is the hormone called oxytocin.

Oxytocin was in fact the hormonal luminary of the conference, coming up repeatedly in discussions of nearly every type of animal bonding: parental, fraternal, sexual and even the capacity to soothe one's self. Dr. C. Sue Carter of the University of Maryland in College Park, one of the organizers of the conference, is renowned in the field of oxytocin research. She suggested in her talk that oxytocin might have played an essential role in the evolution of social behavior, particularly for mammals.

"The neuroendocrinology of lactation may be important to the wiring of the mammalian brain," she said. "Its development was revolutionary."

Oxytocin's first and strongest role may have been in helping to forge the mother-infant bond. But its ability to influence brain circuitry may have been co-opted to serve other affiliative purposes that allowed the formation of alliances and partnerships, thus hastening the evolution of advanced cognitive skills.

Dr. Carter also emphasized that the capacity to affiliate with others increases, not just the quality of life, but its length as well; animals that live in groups enhance one another's chances of survival, not to mention the survival of each other's offspring. "Social behavior contributes to both individual survival and reproductive fitness," she said. "You can have it all."

Dr. Cort A. Pedersen, an oxytocin researcher from the University of North Carolina at Chapel Hill, chimed in with a paean to maternal behavior as the source of the world's brilliance. "Sustained maternal protection and nurturing of offspring until they were able to fend for themselves allowed a much higher rate of survival," he said. "Mothering also permitted a much longer period of brain development and was therefore a prerequisite for the evolution of higher intelligence. Species that mother their offspring have come to dominate every ecological niche in which they dwell."

In his studies, Dr. Pedersen has demonstrated that oxytocin is essential for the initiation of maternal behavior in a rat after giving birth, but that other cues, like the sensation of the pups suckling the mother's nipples or the taste of the pups as the mother licks them, eventually replace the hormone as a sustainer of motherly affections. Consequently, oxytocin levels subside.

Yet if the mother and her pups are kept physically separated in the cage, unable to interact in any way other than visually, the mother's oxytocin concentrations remain elevated. Once reunited with the pups, no matter how many days later, she immediately resumes all her mothering duties. But that willingness to nurture anew is dependent on the sustained oxytocin pulses during the semiseparation; if given a drug that blocks the effects of the hormone, she appears to forget that she ever gave birth at all.

Other scientists at the conference argued that the root of affiliative behavior lies not in motherhood but in the act preceding it: sex. The need for one organism to meld its genes with another drives the need for one organism to overcome temporarily any innate antipathy it may feel toward strangers and cooperate—affiliate—long enough to court and spark.

"Reproduction is the single most important event in an animal's life," said Dr. David Crews of the University of Texas at Austin. "I believe that social and affiliative behaviors evolved from reproductive behaviors." And once such anti-antipathy mechanisms had been established—for example, in the limbic areas of the brain, which control emotion and sex drive— such mechanisms could be put to use for entirely new purposes, like promoting the formation of monogamous pairs to rear offspring or teams of creatures to fend off predators.

Dr. Crews and others pointed out that even blue-green algae, which reproduce asexually by cloning themselves, must for unknown reasons come together en masse before any one of them can beget new buds. Somehow, group life—perhaps chemically, perhaps physically—stimulates their individual photocopying efforts.

In other words, the forces that impel a living creature to seek succor from others of its kind are knit into the very principle of life; even DNA, for that matter, is a double-stranded molecule, pair-bonded with its complementary mate.

To speak of algal affiliation, though, strikes most researchers in the field as stretching the concept of sociality beyond recognition. Dr. Stephen W. Porges of the University of Maryland in College Park sees the ability to affiliate in any meaningful way as a largely mammalian and avian privilege, and he views it as the by-product of mammals' comparatively efficient metabolism. In a novel theory with the faintly mystical name of "the polyvagal theory of emotion," Dr. Porges proposes that the capacity for emotion, and its consequent role in social behavior, is dependent on the advanced nature of the mammalian autonomic nervous system, the part of the nervous system that is essentially automatic, controlling vital functions like heart rate and digestion.

Through phylogenetic comparisons across the evolutionary spectrum, Dr. Porges has identified the vagal nerve complex, a cardinal component of the autonomic nervous system, as a possible key to the development of mammalian emotions and hence sociality. This complex has ancient origins in vertebrate evolution, beginning as a simple connection between the brain stem and the gut, heart and other organs of the body. Its original purpose was to conserve energy. When a fish encounters a reduction of oxygen in its watery world, for example, its primitive vagus nerve slows its heartbeat and digestion and thus cuts back on the fish's oxygen needs.

As life became more complicated and new threats arose, Dr. Porges said, the vagal system of nerve fibers likewise became elaborated. It split into two divisions: the original, oxygen-thrifty circuitry—the dorsal vagal complex—and a newer component that communicates with the sympathetic nervous system. Each element of the vagal nerve complex links up to a different region of the brain stem. Rather than slowing things down, the sympathetic nervous system speeds things up; it allows an animal to use more than the usual amount of oxygen and thus makes possible the famed fight-or-flight response.

With the arrival of mammals came yet a third element of the vagal system, what Dr. Porges calls the "smart" vagus. This vagal nerve complex controls the facial muscles and the larynx, and therefore allows facial expressiveness and vocalizations, the kindred souls of emotionality. It is also coupled with the regulation of the heart, breathing and digestion, keeping the heart beating and digestion running smoothly; at the same time, it inhibits the sympathetic nervous system—the fight-or-flight response—to prevent a state of hyperstimulation that would needlessly burn oxygen and calories. The new vagus allows one to make all sorts of facial expressions—a smile, a frown, an artful widening of the eyes—and any number of calls, or, in the case of people, words, with little effect on breathing or metabolism.

"This enables us to signal other organisms and to be fully engaged in our surroundings, without major metabolic demands or challenges," Dr. Porges said. Yet even with the sophisticated new vagus, the older vagal systems remain, Dr. Porges said, and are called upon should the higher vagal reactions fail. If we try to talk our way out of danger, for instance, and find that the threat remains, the fight-or-flight response kicks in. If we cannot escape no matter how fast our heart beats and muscles throb, and our attacker moves in for the kill, the primitive dorsal-vagal response assumes command as a final attempt to save all life-support systems. It dramatically slows breathing and heart rate, sending one into a state of terrified shock. Alas, the last-ditch effort may prove fatal, Dr. Porges said, for a mammal cannot live long without oxygen coming in and with a heart rate slowed to nearly zero.

In a sense, then, when we lose our capacity to affiliate, we may be on our way to dying of fright.

—NATALIE ANGIER, April 1996